hubble

the mirror on the universe

hubble

the mirror on the universe

David & Charles

Robin Kerrod

A QUINTET BOOK

First published in the UK in 2003
David & Charles,
Newton Abbot
Devon TQ12 4PU

A catalogue record for this book is available from
the British Library

ISBN 0 7153 1642 7

This book was conceived, designed and produced
Quintet Publishing Limited
6 Blundell Street
London N7 9 BH

Project Editor: Toria Leitch
Art Director: Tristan de Lancey
Designer: Roger Fawcett-Tang
Illustrator: Richard Burgess

Managing Editor: Diana Steedman
Creative Director: Richard Dewing
Publisher: Oliver Salzmann

Manufactured in Singapore by Pica Digital Pte Ltd

Printed in China

Visit our website at www.davidandcharles.co.uk

David & Charles books are available from all good
bookshops; alternatively you can contact our
Orderline on (0)1626 334555 or write to us at
FREEPOST EX2 110, David & Charles Direct, Newton
Abbot, TQ12 4ZZ (no stamp required UK mainland).

Contents

Preface

Circling silently in space a few hundred miles above our heads is one of the most amazing scientific instruments ever made — the Hubble Space Telescope (HST). Day after day it is returning images of the universe with incredible beauty, as it peers deep into space to spy on the stars, the vast stellar nurseries where they are born, and distant galaxies.

It was not always so. The HST made a disastrous debut. When first launched in April 1990, it had faulty vision because its primary light-gathering mirror had been inadvertently ground to the wrong shape. The mirror was only two-fiftieths of the width of a human hair out of true, but this was enough to blur the images the HST produced.

But the very perfection of the imperfection of the primary mirror made it relatively easy, optically speaking, to correct. And in December 1993, spacewalking astronauts recovered the HST in orbit and corrected its flawed vision. The telescope then at last began to fulfil its promise of opening up a new window on the universe. On subsequent servicing missions, new instruments have broadened and deepened the view through that window.

The HST is a monster of a satellite, as big as a bus and twice as heavy as a bull elephant. Yet it can be manipulated remotely with microsurgical precision to view a particular region of space. And it can stay locked onto that region for hours at a time, even though it is hurtling in orbit around Earth at more than ten times the speed of a rifle bullet.

Compared with leading ground-based telescopes, the HST has quite a small light-gathering mirror — the Keck telescopes in Hawaii have mirrors more than four times the size. But the HST is a much more effective light-gatherer than ordinary telescopes: they look at the heavens through the "dirty window" that is the Earth's atmosphere, whereas the HST looks at the universe through the pristine clarity of airless space. So it can detect the faintest smudges of light coming from even the remotest depths of the universe. It also has the advantage over ordinary telescopes that it can detect invisible radiations from the heavens that the atmosphere absorbs.

This book reveals the wonderful, mysterious and awesome universe of ours through the eyes of the Hubble Space Telescope. You don't have to be an astronomer to appreciate the often breathtakingly, hauntingly beautiful images, which chronicle frozen moments in the life of the cosmos — from Martian dust storms to extrasolar planetary systems; from the birth pangs of young stars to the death throes of ancient ones; from starbursts in neighboring galaxies to catastrophic collisions in remote galactic clusters.

So travel with me on an incredible journey into space. The ultimate incredible journey. To the ends of the universe. To the beginning of time.

Robin Kerrod
Salisbury, England

left
Intricate jewel
Planetary nebulas have been the subject of many of the most stunning pictures taken by the Hubble Space Telescope. This one is named the Spirograph Nebula for its intricately woven structure, which defies explanation.

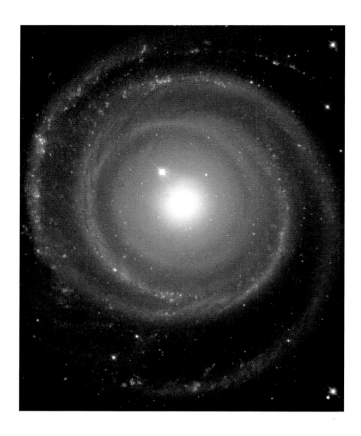

Introduction

Before we look at the universe through the HST's supersensitive eyes, let's set the scene. What, broadly speaking, is this universe of ours like? One thing for sure is that it is vast—unimaginably vast. The Earth, Moon, planets, Sun and stars are nothing but tiny specks of matter floating in an unfathomable immensity of space—miniscule, insignificant plankton floating in an infinitely deep cosmic ocean.

When we cast our eyes up to the night sky and see a glittering firmament of twinkling stars set in the velvety blackness that is space, we are looking out at one little corner of this ocean, of this universe of ours.

Dominating the sky by night is the silvery Moon, Earth's closest companion in space—its only natural satellite. An airless, heavily cratered ball of rock, it is the only other world that human beings have ever set foot on—yet.

Next come the ultrabright stars that seem to wander through the heavens. But they are not stars at all: they are planets. What an extraordinary collection of bodies the planets are. They are all quite different from the planet we know best—planet Earth. Closer to the Sun, Mercury and Venus are oven-hot, while Mars farther out is cold, but might once have been warmer and supported some kind of life. Farther out still are gigantic Jupiter and other gas giants. And farthest out, Pluto is different from them all, a tiny ice world.

Dominating the sky by day is the golden orb of the Sun, which brings warmth, light and life to Earth. It also dominates near space with its powerful gravity, keeping the planets and a host of smaller bodies circling round it. All these bodies make up the Sun's family, or solar system.

The Sun is quite a different body from the planets: it is a huge globe of searing hot gas. It is our local star, just like the myriad other stars in the heavens but very much closer. The other stars, those pinpricks of light in the night sky, are so far away that their light takes years to reach us on Earth (light from the Sun takes just over eight minutes). Astronomers use the distance light travels in a year (about 6 trillion miles/ 9.5 trillion km) as a measure for expressing distances in space. They call it the light-year. It is the unit we use throughout the book.

The Sun is a very ordinary star, of about average size (nearly 1 million miles/1.6 million km across) and average brightness. There are stars that are very much bigger and brighter, and others that are very much smaller and dimmer.

Since the dawn of astronomy at least five millennia ago, stargazers have used patterns made by the bright stars to guide them across the night sky. These patterns are the constellations. Astronomers use Latin names for the constellations, which refer to figures ancient stargazers thought they could see in the patterns of stars. A few constellations live up to their names (Leo, the Lion; Scorpius, the Scorpion; Cygnus, the Swan), but most don't.

All stars are born in great billowing clouds of gas that occupy the space between the stars. After shining steadily for

above
Going backward
The Hubble Space Telescope's sharp eyes are helping astronomers to answer difficult questions such as why this spiral galaxy has arms pointing in opposite directions, and is spinning in the opposite direction from that expected. Hubble's view of the core suggests that an ancient galactic collision might have been responsible for the strange behavior.

perhaps tens of billions of years, stars begin to die. They may exit the universe relatively quietly, as the Sun will eventually, or disappear in a fantastic supernova explosion. The end products of their death throes will be superdense bodies like white dwarfs and neutron stars, or the most awesome objects we know in the universe — black holes.

The stars we see in the night sky may lie many thousands of light-years away, but they are still close neighbors in the universe. They all belong to a great star island in space — a galaxy. The universe is made up of innumerable island galaxies, separated by virtually empty space.

Our galaxy, called the Milky Way or just the Galaxy, probably contains as many as 200 billion stars. It measures some 100,000 light-years across and has a spiral shape. Many galaxies are like it, but others are elliptical in shape or have no regular shape at all. We can see just three galaxies in the night sky with the naked eye. They are the Magellanic Clouds in the far southern skies and the Andromeda Galaxy in northern skies.

Some galaxies are extraordinary, pumping out much more energy into the universe than usual, particularly at radio wavelengths. Called active galaxies, they include enigmatic bodies such as quasars and blazars. Black holes seem to be the engines that generate their exceptional power.

Just as stars group together to form great stellar island galaxies, so galaxies themselves group together to form clusters. Our own Galaxy is part of a relatively small group of about 30 galaxies. But we know of clusters containing thousands of galaxies. In their turn even clusters group together to form superclusters. And on the largest scale, it is strings of superclusters, interspersed with empty voids, that make up the universe.

How can we put such an enormous universe in perspective? With great difficulty — but we can try. Let's suppose we have been able to build an interstellar and intergalactic starship, capable of traveling at the speed of light, and take an incredible journey into space. Setting off from Earth, we would reach the Moon in $1\frac{1}{2}$ seconds and flash pass Venus in $2\frac{1}{2}$ minutes. In less than $8\frac{1}{2}$ minutes we would be leaving the Sun behind, heading for the most distant planet Pluto. We would reach this tiny world in $5\frac{1}{2}$ hours. But it would take several months longer before we escaped completely from the gravitational influence of the Sun and left the solar system. Now traveling in interstellar space, we wouldn't reach even the nearest star (Proxima Centauri) for more than 4 years.

To explore our Galaxy would require flight times measured in tens of thousands of years — 25,000 years to reach the Galaxy's center, twice as long again to reach its edge. To make a visit to our galactic neighbor, the Andromeda Galaxy, we would have a journey time of 2.5 million years. And to reach the farthest objects we can see in the universe, we would have to journey for at least 12 billion years. This is nearly as long as the universe has existed.

1 | Stars in the Firmament

Dark, dusty clouds are the birthplace of stars

above
Cradle of creation
The Omega Nebula (M17) is a hotbed of new-born stars, wrapped in colorful blankets of glowing gas. This was one of the first images taken by the Advanced Camera for Surveys (ACS) in April 2002. The Omega Nebula, also known as the Swan, lies 5,500 light-years away in Sagittarius.

inset left
Seeding interstellar space
Jets of gas stream from an enigmatic object known as He2-90 in Centaurus. It seems to be a close pair of dying stars masquerading as a single youngster—most stars emit jets in their youth. Dying stars add substance to the interstellar medium by shrugging off large amounts of gas and dust.

inset right
Clustering together
A dazzling cluster of stars in one of our galactic neighbors, the Large Magellanic Cloud. The stars in the cluster (NGC 1818) seem to be relatively young, around 40 million years old. Many stars are born in small groups or larger clusters; single stars like the Sun are less common.

The Universe of Gas and Dust

In every direction we look in the night sky, we see stars. If we were very patient and very meticulous, we could count around 3,000 stars visible to the naked eye. Looking through binoculars, we would see thousands more stars, and, through a telescope, stars in their billions.

Even a casual glance reveals that stars are not all the same. Some are bright, others dim; most are white, others colored. Brightness and color are two factors that make stars different. But when we investigate stars closely with telescopes and analyze their faint light, we find that they also differ in many other ways—in temperature, mass, speed, magnetism, composition, age and so on. In particular, other stars can differ remarkably from the star we know best—our local neighborhood star, the Sun.

So many kinds of stars. Tiny stellar dwarfs that glimmer feebly, like celestial glowworms; supergiant stars thousands of times bigger and millions of times more brilliant that stand out like cosmic beacons; stars seemingly with a dimmer switch that makes them vary in brightness; newborn stars with the exuberance of youth; ancient stars racing toward celestial Armageddon.

There are so many different species in the stellar zoo. But in studying these disparate bodies, astronomers discover that each kind represents a different stage of stellar evolution. From this, they can piece together the life cycle of a typical star, from its birth, through its youth, into middle age and finally to its death. Astronomers have to do things this way for they can't follow a single star through all its evolutionary stages, which, of course, would take billions of years.

The story of the stars begins not with the stars themselves, but in the vast space that exists among the stars. We tend to think of this space as being empty, but it isn't—quite. Scattered about interstellar space are atoms of hydrogen gas and tiny grains of dust. On average, there are a few dozen hydrogen atoms and the odd speck of dust in a volume of space the size of a soda can. This means that interstellar space is millions of times more empty than the most perfect vacuum that scientists can achieve on Earth.

The tenuous mixture of gas and dust among the stars is known as the interstellar medium, and its typical composition is called the cosmic abundance of the elements. Hydrogen makes up about 75 percent of the interstellar medium, and helium makes up about 23 percent. The remaining two percent is made up of traces of heavier elements, which astronomers rather confusingly call "metals," though this term also encompasses elements such as carbon and oxygen, which scientists consider non-metals.

Carbon and oxygen are among the most abundant of these "metals," carbon being the primary constituent in interstellar dust grains. Nitrogen and iron are relatively common as well. There are also sprinklings of all the other heavier chemical elements—from arsenic to tin, lead to gold.

Where exactly do all of these elements come from? The hydrogen and helium are primordial—they were formed at the very beginning of the universe. All the other elements, however, have been forged in the interior of stars. Ordinary stars such as the Sun produce carbon and oxygen as they enter old age. Supergiant stars produce iron. But all heavier elements are produced in the supernova explosions that mark the death of supergiants.

All these elements are puffed or blasted into the interstellar medium when a star dies and subtly changes its composition.

above
"Cleft into three"
The Trifid Nebula (M20) in Sagittarius is a glorious object through a telescope. It is named for its prominent dark dust lanes, which divide it into three. Radiation from a hot central star within the nebula makes its gas (hydrogen) glow pink.

right (main image)
Stellar nursery
Dark, brooding clouds of gas and dust in the Trifid Nebula, dramatically lit by intense radiation from a nearby massive star. Over time, the radiation will eat away the cloud and expose any stars that have managed to form within it.

right inset
Interstellar twisters
The HST has spotted all kinds of features in the Lagoon Nebula (M8) in Sagittarius. They include not only glowing clouds and dark globules, but also jets, shock waves, and strange twisted and funnel-shaped features that look somewhat like terrestrial tornadoes.

left
Ponderous pillar
Named for its striking shape, the Cone Nebula in Monoceros is a huge pillar of gas and dust. It is lit up by ultraviolet radiation from young, hot stars behind. This was one of the first images taken by the ACS in April 2002 to demonstrate the new instrument's superlative performance.

right
The horse's head
One of the most spectacular of all dark clouds, the Horsehead Nebula in Orion, also known as Barnard 33. With its uncanny resemblance to a horse's head, the cloud is silhouetted against a bright nebula (IC 434) close to Zeta Orionis, the most southerly star in Orion's Belt.

Molecular Clouds

In many regions of interstellar space, the gas and dust becomes relatively dense, though still nowhere near as dense as air. It then forms into vast cloud-like masses. Typically, these clouds are very still and cold, around minus 430 degrees Fahrenheit (−260°C). Under these conditions, the atoms of the elements present join to form molecules, which is why these cool, dense regions are called molecular clouds.

The most common molecule in the clouds is, of course, hydrogen. Scientists have identified many other interstellar molecules, usually by investigations done at infrared, microwave and radio wavelengths. Every molecule has a characteristic wavelength "signature" that identifies it.

Among the other molecules found in molecular clouds are water, ammonia, hydrogen sulfide, formic acid, methanol and glycine. The presence of glycine is particularly interesting because this compound is an organic substance called an amino acid. And amino acids are the building blocks of proteins, which are essential to life as we know it. So the presence of glycine and the other organic chemicals in interstellar space suggests that there may be forms of life elsewhere in the universe.

The Dark Side

Dark molecular clouds lurk throughout the Galaxy. Most merge into the inky blackness of space. But some become visible when they blot out the light of stars or glowing gas behind them. We call these molecular clouds dark nebulas—from the Latin word for clouds.

We can see two huge dark nebulas with the naked eye. One, called the Cygnus Rift, lies in the constellation Cygnus. The other, appropriately named the Coal Sack, lies in the far southern constellation Crux, the famous Southern Cross. Best-known among dark nebulas visible with telescopes is the Horsehead in the constellation Orion. It really does look just like a horse's head and flowing mane.

All Lit Up

Sometimes the clouds of interstellar gas are lit up by nearby stars and become visible as bright nebulas. They are lit in one of two ways: from the outside or the inside.

When there are stars near the Earth side of a nebula, it reflects their light toward us and is called a reflection nebula. When the stars are embedded in the nebula, radiation from them excites the gas molecules, making them emit their own radiation. We then call it an emission nebula. An emission nebula is characteristically red, as this is the wavelength of light emitted by excited hydrogen atoms.

There are many bright nebulas visible in the heavens. These are predominantly emission nebulas, but they also usually feature smaller regions that glow by reflection. Brightest by far and visible to the naked eye is the Orion Nebula (profiled on page 24).

When you look through a telescope, hundreds of bright nebulas swim into view. Many have distinctive shapes that have earned them popular names, such as the Eagle Nebula in Serpens, the Lagoon Nebula in Sagittarius and the North America Nebula in Cygnus.

following pages
Through the Keyhole
Curving filaments of glowing gas and a lacework of silhouetted cold dark clouds make up the Keyhole Nebula. It forms part of the Eta Carina Nebula— Eta Carinae itself lies just beyond the upper right of the image.

The Orion Nebula

Orion is perhaps the finest constellation in all of the heavens. Spanning the celestial equator, it is visible to stargazers anywhere in the world. Orion is one of the few constellations that bears more than a passing resemblance to the figure it is named for. It depicts a mighty hunter from Greek mythology, in his left hand a shield and in his right a club, raised and ready to strike.

The individual stars that make up Orion are spectacular. The bright orange supergiant star Betelgeuse marks his right shoulder, while brilliant white Rigel marks his left knee. Defining Orion's belt are three slightly less bright stars, and hanging from the belt is his sword, also called the sword handle.

The sword is marked by a bright, misty patch known as the Orion Nebula, or M42. Binoculars reveal greater detail, and with a small telescope we can make out individual stars. But long-exposure photographs in larger instruments are needed to bring out fully the nebula's incredible beauty.

Some 1,600 light-years away, M42 is the nearest bright nebula to Earth. It is what astronomers often call a blister nebula—a region at the edge of a dark molecular cloud that has been lit up by the powerful ultraviolet light of nearby embedded stars.

In M42, the star in question is Theta Orionis. Actually, it is a multiple star system, known as the Trapezium because of the arrangement of its four stars. Only about 300,000 years old, these stars are in their infancy, and they are pumping out vast amounts of energy in the ultraviolet.

But this is not the only region in Orion where star formation is taking place. The whole constellation is embedded in vast, billowing clouds of gas and dust. There are two particularly dense areas, known as the southern and northern molecular clouds.

M42 is part of the southern cloud, which nearly merges into the northern cloud around Zeta Orionis, the most southerly star in Orion's Belt. The outstanding feature of the northern cloud is the Horsehead Nebula, which is thrown into dramatic silhouette by the bright nebula beyond (see page 21).

above
The Sword of Orion
The bright patch we see in the sky as the Sword of Orion reveals itself in telescopes as one of the most beautiful objects in the whole heavens. It is just the visible part of a much bigger dark cloud where intense star formation is taking place.

right
The Trapezium
In this infrared view of the Orion Nebula, the stars of the multiple system we know as Theta Orionis, or the Trapezium, shine brilliantly. The picture also reveals dozens of feebly glowing brown dwarfs—objects too small and too cool to achieve true stardom.

Orion's star factory
This dazzling panorama highlights the brightest
part of the Orion Nebula. It is a mosaic of 15
separate HST images, spanning a region of space
about 2.5 light-years across. Among the features
seen in this turbulent star factory are more than
150 proplyds — solar systems in embryo.

A Star Is Born

The cold, dark molecular clouds found in Orion and elsewhere in our Galaxy can stretch for hundreds of light-years. It is in such giant molecular clouds that stars are born. These clouds can remain relatively still and quiescent for hundreds of millions of years. Then something disturbs them, triggering a chain of events that will ultimately lead them to spawn new generations of stars.

Exactly what triggers a disturbance still puzzles astronomers. It could be the shock wave from a nearby supernova explosion, a collision with another giant molecular cloud or a galactic pressure surge. But the end result is the same: Parts of the cloud become so dense that gravity becomes the most dominant force, and they start to collapse.

Barnard and Bok

The collapsing regions in a molecular cloud may contain enough gas and dust to create hundreds of stars. Such regions are known as Barnard Objects, after the US astronomer Edward Barnard. Typically tens of light-years across, they are the kind of dark nebulas that we often see obscuring the light of background stars. We see other, smaller collapsing regions as little black bubbles against a background of stars or a bright nebula. They are called Bok Globules, after the Dutch astronomer Bart Bok. Only a few light-years across, these contain enough mass for only a few stars.

As both Barnard Objects and Bok Globules collapse, clumps of matter within them collapse as well. So the original cloud fragments successively into smaller and smaller clumps. And it is from the smallest clumps that individual stars form.

Gravity Rules

In the center (core) of a collapsing clump, concentrations of matter build up. The core is a star in embryo—a protostar. The denser the protostar becomes, the stronger is its gravity and the more it attracts surrounding matter. As gravity tugs at it, matter speeds up; and gravitational energy converts into kinetic energy—the energy of motion.

When this moving matter suddenly reaches the core, it collides with the matter already there and stops. Its energy immediately converts into another form—heat. So, as the protostar accumulates more and more material, its temperature rises.

In a Spin

The protostar is not stationary. The original giant molecular cloud rotates, and it imparts this rotation to the collapsing clump and the protostar within. The spinning mass then flattens out. Over time a thick disk, or ring, of gas and dust builds up around the protostar, while its temperature rises rapidly. It is already giving off radiation, mainly at infrared wavelengths.

As more and more matter rains down, temperatures inside the protostar soar to millions of degrees, and pressures reach millions of atmospheres. Under these exotic conditions, the nuclei (centers) of hydrogen atoms are forced together, and they fuse (join up). Nuclear fusion produces fantastic amounts of energy (see page 118). The protostar springs to life and begins pouring out light, heat and other radiation into the universe. A new star is born.

above
Colorful jets
At the center of this HST image of Herbig-Haro object 32 (HH 32) is a young star blasting jets of matter into space in polar jets. One jet (at the top) is seen plowing into interstellar gas and making it glow in the light of hydrogen atoms (green) and sulfur ions (blue). The jet streaming in the opposite direction is mostly obscured by dust.

left
Thackeray's globules
Dark clouds float serenely against a backdrop of light from a loose cluster of young, massive stars in a stellar nursery known as IC 2944. Cool and dense, these dust clouds are named Thackeray's globules for the astronomer (A. D. Thackeray) who first spied them in 1950.

below
Enigmatic "waterfall"
Both polar jets from
a protostar can be
seen slamming into
interstellar gas in
this VLT (Very Large
Telescope) image of
HH 34. The nature of
the prominent waterfall-
like stream of light is
a mystery.

right
Double bubble
A cocoon of gas and
dust surrounds a small
cluster of young hot
stars in the Large
Megallanic Cloud.
This "double bubble"
lies inside the larger
nebula, DEM L106, one
of many star-forming
regions in the galaxy.

Reaching Equilibrium

Once the nuclear furnace of a star fires up, radiation floods out of it. Until this point, the star has been contracting under gravity. But now radiation pressure starts pushing in the other direction, causing the star to expand. It continues to expand until it reaches an equilibrium point, at which the outward push of the radiation balances the inward pull of gravity. At that point, the newborn star reaches the steady state it will assume for many millions — or even billions — of years. It becomes what astronomers call a main sequence star (see page 35).

In reality, however, things are not that simple. It takes a star quite a while to reach the equilibrium point, typically about 10 million years. When it first starts to expand, battling against gravity, it expands too far and blows material off into space. Then gravity takes charge for a while, forcing the star to contract again. Soon the cycle starts all over again — expansion, mass loss, contraction; expansion, mass loss, contraction. Slowly, the amplitude of these oscillations subsides, until the star finally achieves a state of equilibrium.

The cycle of expansion and contraction causes the star's output of light to fluctuate erratically. This period of a star's life is called the T Tauri phase, after the first star of its kind that was discovered.

above and right
Jet exhausts
Jets are common "exhaust" products of star formation, revealed when they ram into surrounding gas and dust. In these images, the central stars are hidden within masses of infalling material. Typically, the twin jets from a fledgling star span a region of space about two light-years across.

left
Making waves
Gas streaming from the very young star LL Ori collides violently with the tenuous interstellar medium, creating a "bow shock" around it.

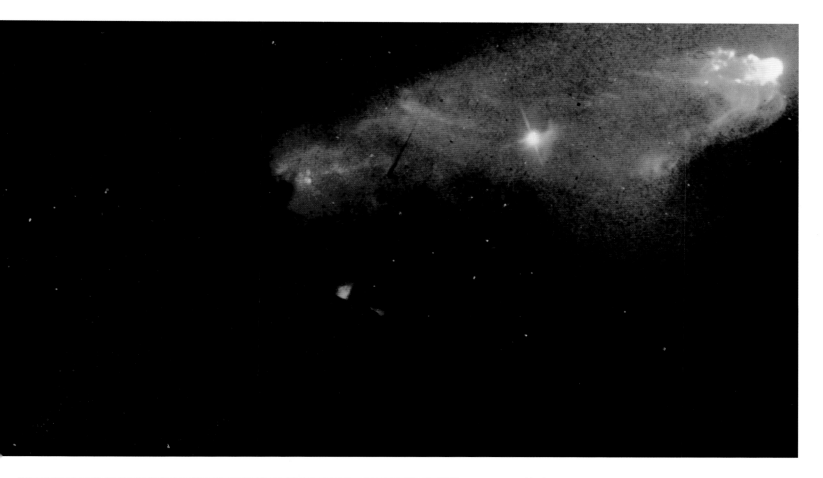

Jet-propelled

In these early days of a star's life, when its nuclear furnace is stoking up, it is still surrounded by a veil of gas and dust and a thick disk of material. It also gives off streams of particles as a kind of stellar "wind" that starts to blow away the surrounding material. This can't happen around the equatorial regions because of the surrounding disk. But it can and does happen at the poles, where matter emerges as powerful jets. The jets stream out into space and collide with interstellar gases, creating "knots" of glowing gas known as Herbig-Haro objects. The jet phase of a star's life is fleeting, lasting only a few thousand years.

CoKu Tau1

IRAS 04248+2612

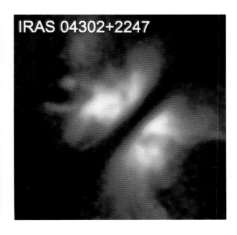

IRAS 04302+2247

left
Planets in the making
Dark, dusty disks around embryonic stars show up in these infrared images. When the stars begin shining steadily, the disks will probably remain, providing raw materials for future planetary systems.

The Pillars of Creation

Serpens (the Serpent) is an oddity among the constellations because it is split in two: Serpens Caput—the serpent's head—and Serpens Cauda—the serpent's tail. In between is the constellation Ophiuchus, the Serpent-Bearer.

Ophiuchus, like Serpens itself, is not an easy constellation to identify because it has no particularly bright stars. One of its claims to fame is what it's not—it's not considered by astrologers to be a constellation of the zodiac. And it should be, because the Sun spends longer passing through Ophiuchus than it does passing through Scorpius, which is a zodiac constellation. The omission of Ophiuchus as a star sign (or zodiac constellation) is just one serious weakness in the case for astrology.

But, back to Serpens. Serpens Caput has a fine globular cluster, M15. Serpens Cauda has M16, which is a bright nebula vaguely shaped like a bird with outspread wings. Known as the Eagle Nebula, M16 was the subject of one of the most dramatic images the Hubble Space Telescope (HST) has ever taken, on April 1, 1995.

The HST science team called the image "The pillars of creation." It shows dark columns of gas in which stars are being born. The columns, or pillars, are etched and silhouetted by the light of young, hot, massive stars beyond. The pillar on the left is about one light-year long.

The finger-like protrusions at the top of the pillars are dense regions that probably contain newborn stars or protostars. Termed EGGs (evaporating gaseous globules), they have been revealed because intense ultraviolet radiation from hidden massive stars has blown less dense gas away. Eventually the radiation will blow away the gas in the EGGs as well, revealing the star inside for the first time.

left
Eagle's EGGs
This close-up of the tallest of the pillars in the main picture clearly shows the finger-like globules that astronomers call EGGs. They are about as wide as our solar system.

above
Embedded cluster
This telescopic view
shows the whole Eagle
Nebula and the bright
open cluster embedded
in it. The "pillars" are
found near the center
of the glowing cloud.
Just beyond the limit of
naked-eye visibility, the
nebula is easily spotted
through binoculars.

left
Stellar Stalagmites
The eerie pillars of dark,
dense gas in the heart
of the Eagle Nebula are
incubators for new stars.
They protrude from
the wall of a dark
molecular cloud rather
like stalagmites from
a cave floor.

What Stars Are Like

When a molecular cloud collapses, the stars that form can vary widely in mass. A star's mass determines how long it will shine. Many stars have a mass similar to that of the Sun, which is a relatively small star. But even more stars have a smaller mass — some with less than a tenth of that of the Sun. Stars with much smaller mass don't exist because they don't accumulate enough mass to generate the heat and pressure necessary to trigger nuclear fusion. These "failed stars" glow a dull brown and we call them brown dwarfs.

A collapsing cloud tends to produce relatively few really big stars. But they can be huge, with up to 50 times the mass of the Sun and 20 times its diameter. Bigger stars don't exist because they generate so much energy at the protostar stage that they blow themselves apart.

Mass and Temperature

The mass of a main sequence star determines its temperature. The smallest dwarf stars "burn" their hydrogen fuel relatively slowly and have a low surface temperature, maybe below 5,500 degrees Fahrenheit (3,000°C). Somewhat larger stars like the Sun are about twice as hot, while the most massive giant stars have temperatures beyond 55,000 degrees Fahrenheit (30,000°C).

Allied to a star's temperature is its brightness. Here we are talking about a star's true, or absolute, brightness, not its apparent brightness as we perceive it from Earth. As you might expect, the smallest stars are the dimmest — they may be only 1/100,000,000th as bright as the Sun. And the biggest stars are the brightest — up to 100,000 times brighter than the Sun. The brightness of a star compared to that of the Sun is called its luminosity.

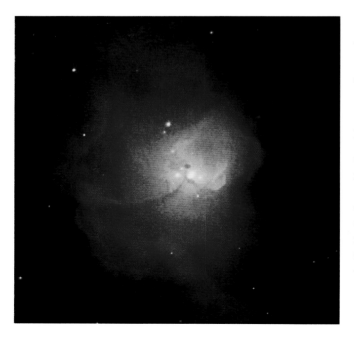

Temperature and Color

The temperature of a star also determines its color. Think of an iron plunged into a fire: As its temperature rises, it gradually changes color from dull to bright red, then orange, yellow and finally brilliant white. And it is a similar story with stars.

Small, cool stars glow red; we call them red dwarfs, or sub-dwarfs if they are very small. Larger, hotter stars like the Sun glow yellow; we call them yellow dwarfs. The biggest, hottest stars glow blue-white; we call them blue giants.

The Informative Spectrum

When we see a rainbow in the sky, we are looking at a range of different colors, from violet to red. Mixed together, these colors make up white sunlight. The spread of color, or spectrum, is a kind of fanning out of the different wavelengths in light because we perceive different wavelengths as different colors. Violet light has the shortest wavelength, red the longest. Scientists create a spectrum artificially by passing sunlight through an instrument called a spectroscope or spectrograph. The technique is called spectroscopy. The light from all stars can be split into a spectrum in a similar way.

Spectroscopy is one of the most powerful tools in astronomy, since a star's spectrum holds the key to its identity. When you closely examine the rainbow-like spectrum of any star (including the Sun), you notice a series of dark lines. These are called absorption lines, because they are produced when certain wavelengths of starlight are absorbed as they pass through the star's atmosphere. Where these lines appear in the spectrum depends on what elements are present in the atmosphere.

Every element absorbs different wavelengths and gives rise to a characteristic set of spectral lines. So by studying the spectrum, astronomers can work out the composition of a star and much more besides. They can also estimate the star's temperature, density and magnetism; tell how fast it is spinning; and determine whether it is traveling toward or away from us.

Be a Fine Guy

All stars produce a spectrum with dark absorption lines, but each kind of star displays a slightly different and characteristic spectrum, depending on its color (and thus its temperature).

By this means, astronomers divide stars into 10 main groups, or spectral classes, designated O, B, A, F, G, K, M, R, N and S. (Difficult to remember? Try the once-heard, never-forgotten mnemonic "Oh, Be a Fine Guy, Kiss Me Right Now, Sweetie!") O and B stars are the hottest; S stars are the coolest. The cooler the star, the more lines appear in its spectrum.

left
Firestorm in the SMC
A cluster of about 50 young, hot massive stars forms the bright core of a nebula (N81) in the Small Magellanic Cloud (SMC), a close neighboring galaxy. Each star blazes with the brilliance of 300,000 Suns.

right
Sagittarius's jewels
Peering into the Sagittarius star cloud (SGR-1), the HST reveals a dazzling array of stars, flashing in all colors like sparkling jewels. SGR-1 is a narrow, dust-free region in the Milky Way, looking toward the heart of our Galaxy.

above
Hot stuff
A Wolf-Rayet star (WR124) ejects vast amounts of matter into space at speeds of more than 100,000 mph (160,000 km/h). Wolf-Rayet stars are a comparatively rare class of stars that are short-lived and super-hot. WR124 has a surface temperature of 90,000°F (50,000°C), nearly ten times hotter than the Sun.

right
Really cool
The main star in the image is the K-class dwarf star Gliese (GL) 105A, which has a surface temperature around 7,000°F (4,000°C). The small star is its faint binary companion, really cool at only about 4,200°F (2,300°C). It is one of the least massive and coolest stars known.

The H-R Diagram

With such a variety of different stars—bright, dim; hot, cool; blue, red; large, small—how is it possible to classify them? To answer this question, we must go back to two of the leading astronomers from the early part of last century: Ejner Hertzsprung of Denmark and Henry Norris Russell of the United States.

Working independently, they discovered that there is a significant relationship between the true brightness (absolute magnitude) of a star and its spectral class/temperature. When they plotted true brightness against spectral class for a variety of stars, they produced a very interesting graph, which is now called the Hertzsprung-Russell (H-R) diagram.

On the Main Sequence

On the H-R diagram, stars are not scattered about randomly, as you might expect. Instead, they fall into a number of fairly distinct groups. The majority lie within an elongated, S-shaped, diagonal band called the main sequence.

Dim, cool red dwarfs, such as Barnard's Star and Proxima Centauri, are found on the lower right of the main sequence. Bright, hot blue giants, such as Spica and Regulus, are found on the upper left. The Sun, a yellow dwarf, is nearly halfway up the main sequence. In effect, the position of a star on the main sequence depends on its mass. The more massive a star, the hotter it is, because it burns more fuel (hydrogen), faster.

Cool Giants and Hot Dwarfs

Stars are also found in three other main regions on the H-R diagram. Giant stars like Aldebaran and Arcturus are found above the main sequence; they are bright but cool. Supergiants such as Canopus and Antares lie near the top of the diagram; they may be hot and blue or cool and red. The very hot but very tiny bodies we call white dwarfs are at the foot of the diagram.

Life Cycle

The H-R diagram reflects a frozen moment in the lives of stars we can see in the heavens today. If in a few billion years time we were to plot an H-R diagram again, it would look completely different.

Stars remain on the main sequence for most of their productive lives—that is, when they are burning hydrogen in their cores. But not all stars are the same. Red and yellow dwarf stars remain on the main sequence for billions of years, but hot blue giants pay only a fleeting visit, measured in a few million years.

When stars have used up all their hydrogen, they begin to die and leave the main sequence. The biggest ones become supergiants, the smallest, red giants. Eventually, supergiants blow themselves to bits and disappear from the diagram; some even disappear from the visible universe (as black holes). Red giants shrink and become very hot, turning into white dwarfs. We follow the often spectacular death throes of stars in the next chapter.

above
Dying dwarfs
White dwarfs (circled) show up faintly in the M4 globular cluster among the other brighter main-sequence stars. M4 is located close to Antares, the brightest star in Scorpius. It is one of the nearest globular clusters, about 7,000 light-years away.

below
The H-R Diagram
A selection of familiar stars plotted on an H-R diagram. They tend to fall into four main groups. Stars in the prime of life ride the main sequence. Giant and supergiant stars lie above this diagonal band, white dwarfs below.

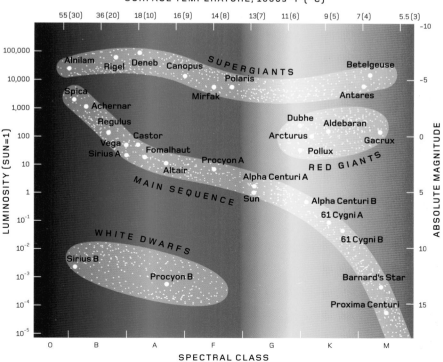

Clustering Together

Stars are born in the giant molecular clouds that exist between the stars, but they are rarely born one at a time. Usually, many clumps of matter condense within a cloud simultaneously, and each one may develop into a star. And if two or more stars are born close together, they may eventually travel as companions through space.

The Sun, of course, has no stellar companion and moves alone in space — except for its family of planets and other bodies. Roughly a third of stars lead solitary lives, but the rest travel with one or more "sibling" stars.

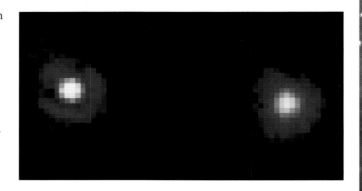

Binaries

Nearly half of all stars are double stars called binaries. In these systems, the two component stars orbit one another in a perpetual celestial waltz. Sometimes we can separate the components — see them separately — in a telescope. But often the components are so close together that we can identify them only by the shifts in their spectral lines as they move to and fro. In this case they are called spectroscopic binaries.

Binaries are not to be confused with optical doubles. These are pairs of stars that are actually far apart, but appear close together only because they happen to lie in the same direction in space as viewed from Earth.

The Winking Demon

A particularly interesting kind of binary is called an eclipsing binary. In this system, the two component stars circle round each other in our line of sight, and each star periodically passes in front of, or eclipses, the other. The overall brightness of the system dips briefly with each eclipse, before regaining its former brilliance when the eclipse is over. Eclipsing binaries are a common kind of variable star — one that varies in brightness. The star Algol ("the winking demon") in Perseus is

a classic example. Other kinds of variable stars, such as Cepheids, vary in brightness because of processes going on inside them (see page 97)

Open Clusters

Sometimes hundreds of stars are born close together and become companions — at least for a while. There are many groups of stars like this in the heavens, called open clusters because the individual stars within them lie relatively far apart.

We can see several open clusters with the naked eye. The easiest to spot is the Pleiades in the constellation Taurus. It is also called the Seven Sisters, because keen-sighted people may be able to see its seven brightest stars. There is also another open cluster in Taurus, called the Hyades. It surrounds (but is not part of) the noticeably orange star Aldebaran, which marks the eye of the bull.

Young Upstarts

Most open clusters are made up of young, hot stars. The Pleiades stars were born about 75 million years ago; those in the Jewel Box in Crux, only about seven million years ago. These clusters are the new kids on the block as far as the universe is concerned, but they will not always stay together. Their gravitational ties are tenuous, and over time they will disperse and gain their independence.

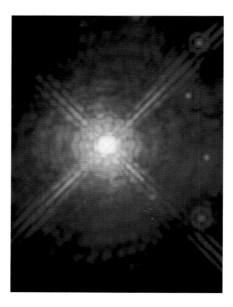

left
Family portrait
This infrared image shows a stellar family of a bright, massive mother star, surrounded by six fainter infants. Violent stellar winds released during the formative years of the mother star triggered star formation in the surrounding gas clouds.

above
Mira and companion
The cool red giant star Omicron Ceti, or Mira (right), is the archetypal long-period variable star, which changes brightness dramatically month by month. It has a close, hot binary companion (left), which can't easily be resolved in telescopes on the ground. The HST does so easily.

right
Ripples in Tarantula
The star cluster in the image, Hodge 301, is located at the edge of the Tarantula Nebula in the Large Magellanic Cloud. It contains a mix of brilliant, massive stars, including several red supergiants that must be close to exploding as supernovas. Other stars in the cluster have already exploded, sending shock waves rippling through the nebula.

Globulars

In the constellation Hercules is an object designated M13. It is just visible to the naked eye, and clearly seen in binoculars. But a powerful telescope is needed to show it in detail. It proves to be a huge mass of stars packed tightly together into a globe shape and numbered in their hundreds of thousands.

M13 is a concentrated group of stars called a globular cluster. It is not unique. More than 100 globular clusters have been found in our Galaxy, and many more exist in others. The two most spectacular globular clusters in our Galaxy are found in far southern skies, and both are easily seen with the naked eye. Looking rather like hazy stars, they are so identified, as Omega Centauri and 47 Tucanae.

Omega Centauri, in the constellation Centaurus, is thought to contain at least a million stars. It is the biggest globular we know, measuring 180 light-years across. 47 Tucanae, in Tucana, may well have as many stars; it is slightly smaller but has a more concentrated central region.

In the Center

Open clusters like the Pleiades are made up of young, hot stars and are found in the spiral arms of our Galaxy (see page 36). Some of these clusters lie quite close to us.

Globular clusters, however, are made up of old stars and are found toward the center of the Galaxy. They do not take part in the Galaxy's general rotation, but pursue independent

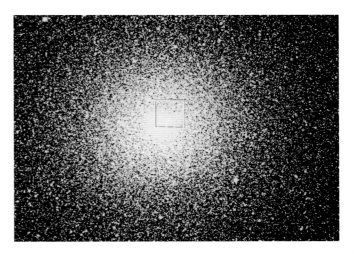

orbits around the galactic center. This takes them well above and below the plane of our Galaxy into the spherical region known as the halo.

Whereas the young and hot stars in an open cluster usually appear blue, globular cluster stars appear yellow because they are older and cooler. Astronomers reckon that globulars are typically about 10 billion years old, so they must have formed during or shortly after the formation of the Galaxy itself.

Studying these ancient stellar bundles helps astronomers to chart the early history of our Galaxy. The HST is playing a particularly crucial role in this study, because it can resolve (view separately) individual stars in the very heart of the globulars. It can also spot globulars in other galaxies, some of which have proved to be much younger than those in our own.

left
Blue stragglers
This dazzling image shows M80, a globular cluster in Scorpius. It lies nearly twice as far away as 47 Tucanae and contains fewer stars. But it contains a large population of "blue stragglers," which appear to be younger, bluer and more massive than the other stars. It is thought that these bodies have formed from the merger of two colliding stars.

above
The Toucan's gem
A ground-based image of one of the astronomical highlights of the Southern Hemisphere, the globular cluster 47 Tucanae. Containing as many as a million stars packed tightly together, it lies about 15,000 light-years away.

left
In the center
This HST view shows about 35,000 stars in the central region of 47 Tucanae. Most of the stars in the cluster formed about 10 billion years ago.

2 | Stellar Death and Destru

Dying stars exit the universe spectacularly

Blowing bubbles
The extraordinary Bubble Nebula in Cassiopeia. The
expanding spherical surface marks the boundary where
particle winds from the central star push up against
surrounding gas. The star is huge and destined to have
a very brief life.

Cometary knots
Tadpole-shaped blobs stream from the inner edge of the
Helix Nebula in Aquarius. They are called cometary knots,
though they have nothing to do with comets. The Helix
Nebula is a close planetary nebula, with gas puffed out
by a dying star.

Sheer beauty
The HST reveals intricate detail and often sheer beauty in
planetary nebulas like this one in Cygnus. Faint concentric
rings show where the central star has ejected wave after
wave of gaseous material.

The Beginning of the End

Stars form out of giant clouds of interstellar gas and dust. Then, with their nuclear furnaces stoked up, they join the multitude of other stars on the main sequence. The Sun has spent around 4.6 billion years on the main sequence shining steadily, pouring out light, heat and other radiation into space. So, will it carry on shining like it does today forever? The answer is: No.

Like living things on Earth, the Sun and all the other stars have a natural life-span. Whereas the life-spans of even the oldest living things, such as California's majestic redwoods, are measured in only a few thousand years, those of stars can be measured in tens of billions of years.

The exact life-span of a star depends primarily on its mass. The bigger the star, the shorter is its life. The Sun is a relatively small star, a type known as a yellow dwarf, which has a relatively long life-span of about nine to 10 billion years. This means that the Sun is now comfortably in middle age. It should carry on shining steadily like it does today for roughly another five billion years. But then, inexorably, it will begin to die.

The Sun and other similar dwarf stars die relatively quietly— we might say, with a whimper. But for stars with much greater mass it is a different story. They depart the celestial scene literally with a bang—the biggest bang in the universe.

A star on the main sequence produces energy by transforming hydrogen into helium in its core. A star like the Sun uses up around four million tons of hydrogen every second, but it is so big that it has enough to last for upward of 10 billion years.

Eventually, though, the hydrogen in the core of the star runs out, and only helium remains. Nuclear fusion ceases. This heralds an abrupt change in the star's life. It is the beginning of the end—the star is dying.

With no radiation emanating from the star's core, gravity once again becomes the dominant force. The core begins to collapse. The potential energy released by this collapse does two things: it increases the temperature and pressure in the core, and it heats up the outer atmosphere of the star.

Cool Giant

The atmosphere expands greatly, making the star swell up to 30, 50 or perhaps even 100 times its original diameter. Because the star now emits radiation from a much greater surface area, its surface temperature plummets by about half, to around only 5,500 degrees Fahrenheit (3,000°C). This lower temperature makes the star produce a redder light. It becomes a red giant and leaves the main sequence behind (see page 35). When the Sun becomes a red giant in around five billion years time, it will expand 30 times or more and become 1,000 times brighter. It will probably swallow up the planet Mercury and maybe even Venus. The Earth will lose its atmosphere, and the oceans will boil away. Life will be eradicated, and our now-verdant home will become a barren cinder of a planet.

left
Red Sun
Sometimes at sunset, white sunlight turns orange and red as it passes through the lower, dustiest part of the atmosphere. In billions of years, the Sun itself will turn red as it swells up to become a giant star.

below

The generation game

In the giant nebula NGC 3603, the HST has captured stars at different stages in their life-cycles. At upper right are dark clouds that are Bok globules, marking an early stage of star formation. In the center is a cluster of youthful hot stars. Above it is an aging blue supergiant on the brink of destroying itself in a supernova explosion.

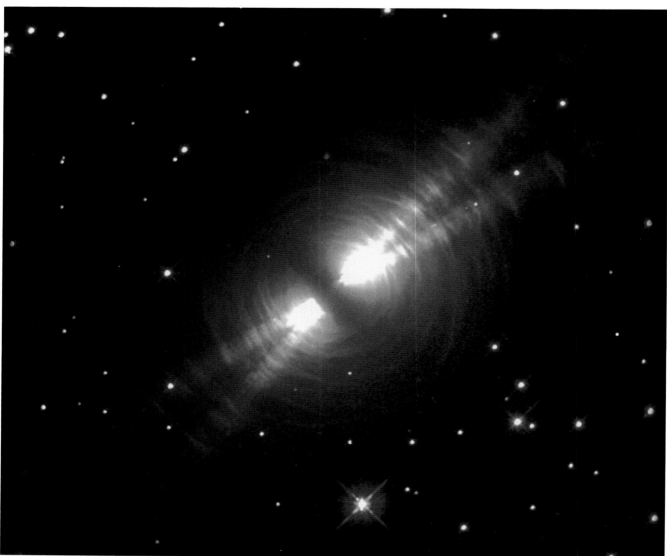

left
Dying Egg
Hidden inside this cocoon of dusty gas (the Egg Nebula) is a red giant star. It has been ejecting waves of material for only a few hundred years as it periodically pulsates.

below
In the infrared
Peering inside the Egg Nebula in infrared light, the HST reveals that the dying giant ejects material along preferred axes, in four broad jets.

Spotting Giants

Some of our most familiar stars are red giants, which appear in the sky with a noticeable orange-red hue. Arcturus, in the constellation Boötes (the Herdsman), is a red giant. It is the most brilliant star in the Northern Hemisphere and the fourth-brightest in all of the heavens. You can easily find it by following the curve in the handle of the Big Dipper, the most recognizable part of the constellation Ursa Major (the Great Bear). Capella in Auriga (the Charioteer) and Aldebaran in Taurus (the Bull) are two other northern giants. Aldebaran marks the baleful red eye of the bull, which appears to be charging the mighty hunter Orion in the adjacent constellation.

In the Southern Hemisphere, Gacrux, one of the four bright stars that make up Crux (the Southern Cross), is a red giant. Its warm orange color contrasts with the other three stars in the cross, which are white.

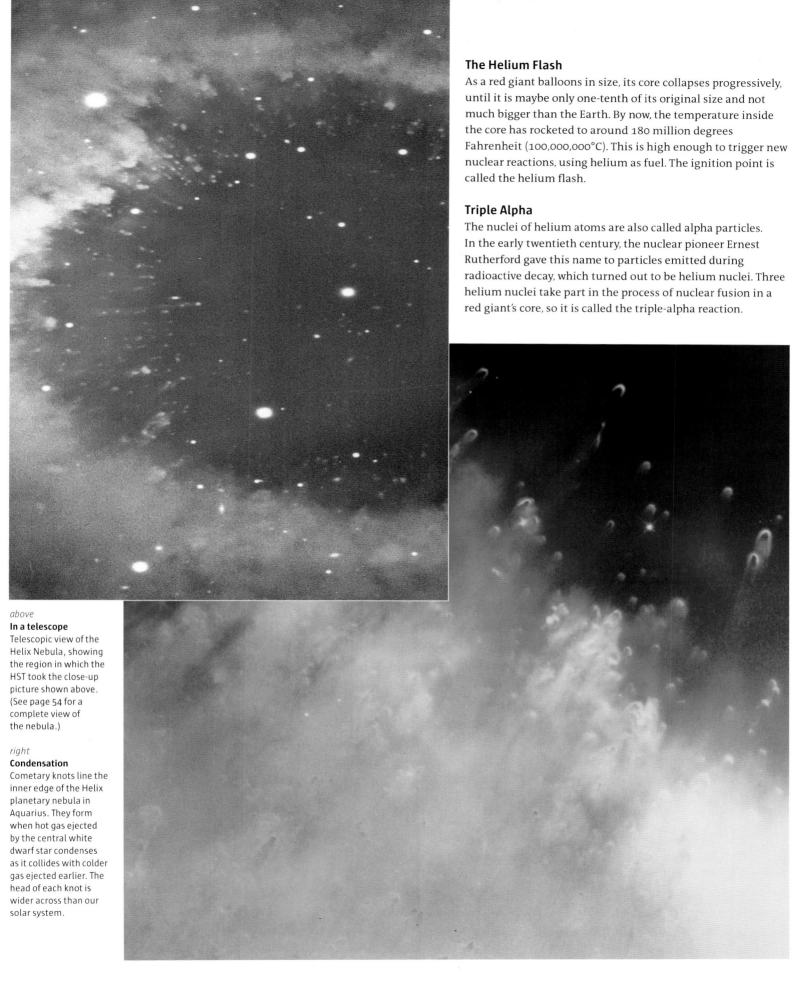

The Helium Flash

As a red giant balloons in size, its core collapses progressively, until it is maybe only one-tenth of its original size and not much bigger than the Earth. By now, the temperature inside the core has rocketed to around 180 million degrees Fahrenheit (100,000,000°C). This is high enough to trigger new nuclear reactions, using helium as fuel. The ignition point is called the helium flash.

Triple Alpha

The nuclei of helium atoms are also called alpha particles. In the early twentieth century, the nuclear pioneer Ernest Rutherford gave this name to particles emitted during radioactive decay, which turned out to be helium nuclei. Three helium nuclei take part in the process of nuclear fusion in a red giant's core, so it is called the triple-alpha reaction.

above
In a telescope
Telescopic view of the Helix Nebula, showing the region in which the HST took the close-up picture shown above. (See page 54 for a complete view of the nebula.)

right
Condensation
Cometary knots line the inner edge of the Helix planetary nebula in Aquarius. They form when hot gas ejected by the central white dwarf star condenses as it collides with colder gas ejected earlier. The head of each knot is wider across than our solar system.

In the process, first two helium nuclei fuse to form an unstable beryllium nucleus. If a third helium nucleus collides and fuses with that, it produces carbon. If a fourth collides and fuses, it produces oxygen.

Although all the hydrogen has been used up in the core, there is still hydrogen in the outer atmosphere. Temperatures are high enough immediately outside the core to cause the hydrogen to fuse. So the core takes on a kind of layered structure: in the center is accumulated carbon and oxygen; around that is an inner shell in which helium is fusing; and surrounding that is an outer shell of fusing hydrogen.

Stellar Winds

The heat produced in the core of a red giant rises to the surface on convection currents. These currents also carry a certain amount of core material, such as carbon, to the surface. The carbon condenses (turns into a solid) in the outermost layers, forming a soot-like dust.

As time goes by, gases in the outer atmosphere of the star gradually stream off into space as a stellar wind, just like the solar wind from the Sun. The star's pulsating—periodic expanding and contracting—sometimes exaggerates the process. The pulsations cause it to vary in brightness, becoming brighter as it contracts and heats up, dimmer as it expands and cools down.

The star Omicron Ceti in Cetus (the Whale) is a classic example of a pulsating red giant. At its brightest, it is second magnitude and is easily visible with the naked eye. At its dimmest, it has a magnitude of 10 and cannot be spotted even with binoculars. It varies from bright to dim, and back again, over a period of about 330 days. The star is usually known as Mira, which means "the wonderful." And the class of long-period variable stars it typifies are called Mira variables.

Dusty Eruptions

Usually, the stellar wind also carries with it the sooty dust grains in the outer atmosphere. Occasionally, however, the dust accumulates around the star and dims its light. Then a vigorous blast of stellar wind removes it, and the star resumes its former brightness. This happens with supergiants as well.

When we observe such a star, we see it as a variable with an irregular period, because the process of build-up and subsequent blowout can take weeks or even years. R Corona Borealis is a classic example. Most of the time it shines at a steady magnitude of six or so, on the edge of naked-eye visibility. But, unpredictably, it can suddenly dim to a magnitude of 12 or below. It may start to recover almost immediately or do so over the course of several months.

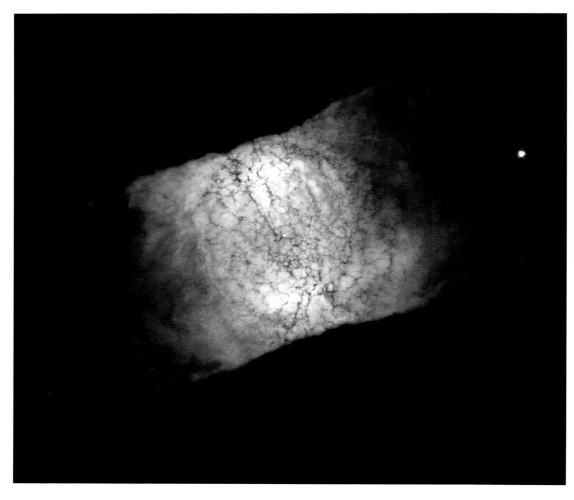

left
The Retina
Glowing in all colors of the rainbow, this planetary nebula (IC 4406) has a donut shape. It has been dubbed the Retina Nebula because the intricate tendrils of material we see have been compared with those in the eye's retina.

right
Hot dwarf
Over 500 times wider than the solar system, this planetary nebula (NGC 6751) resembles a giant eye, with multicolored iris. The temperature of the white-dwarf star at the center could be as high as 250,000°F (140,000°C).

Last Gasps

Helium fusion in the core of a red giant lasts on average about two billion years. Then the helium runs out. Temperatures never get high enough to start fusing the carbon and oxygen that has been produced. In a star like the Sun, this marks the end of energy creation. The star is close to death.

Now come the final death throes. The super-hot core gives off a succession of shuddering death rattles that eject matter into space from the giant's outer layers. Traveling at velocities of up to 20 miles (30 km) per second, the successive waves of matter form into expanding shells. A torrent of X-rays and ultraviolet radiation streaming from the exposed core excite the ejected gas, and the shells light up with Technicolor brilliance.

Planetary Nebulas

Viewed through a telescope, these expanding shells are some of the most beautiful objects in the heavens. They are called planetary nebulas, although the term is misleading because they have nothing to do with planets. Eighteenth-century English astronomer William Herschel gave these objects their name because in the telescopes of the day they looked rather like the disks of planets.

Planetary nebulas take on a number of different forms. Some have the appearance of single or multiple spherical bubbles, like M27, the Dumbbell Nebula in the constellation Vulpecula (the Fox). We see the bubble-like shells of other planetary nebulas as rings, such as the famous Ring Nebula in Lyra (the Lyre), profiled on page 54.

Of Butterflies and Cats

Most nebulas, however, are quite complex in shape. Many have overlapping rings of contrasting colors, caused by successive waves of ejected material punching through earlier ejecta. Others have an hourglass shape, with the ejected gas forming lobes on either side of the dying star. The Butterfly Nebula in Ophiuchus (the Serpent-Bearer) is a beautiful example.

The Cat's-Eye Nebula in the constellation Draco (the Dragon) has an amazingly intricate structure and thoroughly deserves its name. Astronomers reckon that the dying star at its center might be part of a binary system, and that the gravity of the other (unseen) orbiting star has attracted material to create this glorious object.

With its extraordinary ability to distinguish detail, the HST has revolutionized our study of planetary nebulas. It reveals within these stellar ghosts a previously unsuspected complexity of structures and dynamic interactions.

left
The Cat's Eye
Well named, the Cat's Eye Nebula is one of the most beautiful and most complex planetary nebula the HST has spied. Astronomers reckon that the central star could be a close binary system, which might account for the nebula's unusual structure.

left
The Eskimo
The HST's exquisite rendering of the Eskimo planetary nebula, so called because when viewed in ground-based telescopes it resembles a face surrounded by a fur parka.

above
The Southern Crab
This planetary nebula in the far southern constellation Centaurus is aptly named the Crab.

following pages
Southern Ring
Planetary nebula NGC 3132, in Vela, is well named the Southern Ring. It rivals the northern Ring Nebula in Lyra in beauty. The white dwarf ejecting material is the fainter of the two stars in the center.

Heavyweight Dwarfs

The outflow of matter in stellar winds and pulsations can deplete a red giant of as much as 80 percent of its mass. In tens of thousands of years, the nebula will disperse into the general interstellar medium and disappear. All that will then remain is the exposed core.

Once the core runs out of helium, no energy-producing processes can occur there. Temperatures inside are not high enough to trigger a new round of nuclear-fusion reactions with the carbon and oxygen that make up the core.

With no radiation, the core can no longer resist the inward pull of gravity, so it progressively collapses. Eventually it shrinks to about the same size as the Earth and becomes a class of star we call a white dwarf.

The gravitational energy released as the core shrinks manifests itself as heat. The temperature of the surface of the white dwarf soars to tens of thousands of degrees and beyond. The HST has spotted a white dwarf in the planetary nebula NGC 2440 with a surface temperature of 360,000 degrees Fahrenheit (200,000°C). It is one of the hottest stars we know, nearly 40 times hotter than the Sun.

Degenerate Matter

With the mass of the Sun squeezed into a volume the size of the Earth, a white dwarf is extremely dense. Just a teaspoonful of its matter would weigh tons.

Furthermore, the matter that makes up a white dwarf is not ordinary matter. Ordinary matter consists of atoms made up of nuclei, with swarms of electrons circling round them at a distance. In a white dwarf, the matter is crushed. The electrons are practically pressed up against the nuclei. The repulsion between the electrons, which are all negatively charged, prevents the body from collapsing further. This kind of matter is called electron-degenerate matter.

Chandrasekhar's Limit

Typically, the core of a white dwarf has about the same mass as the Sun. It can never have more than 1.4 times the mass of the Sun—a surprising discovery made in 1930 by Indian astronomer Subrahmanyan Chandrasekhar. If the core of a dying star has more than 1.4 solar masses (called the Chandrasekhar limit), gravity overcomes the electron repulsion and causes further collapse (see page 62).

The Final Curtain

What eventually happens to a white dwarf? Most of them just quietly fade away. As they radiate away their remaining energy, they cool and dim, and their light reddens. After many billions of years they run out of energy completely. They become black dwarfs and disappear from the visible universe. Scientists are uncertain whether any black dwarfs exist. It may be that the universe is not yet old enough.

The Pup

The first white dwarf that astronomers discovered was the so-called Companion of Sirius, the Dog Star. It is nicknamed the Pup. (Astronomers call it Sirius B, and the Dog Star Sirius A.) In 1834, the German astronomer Friedrich Bessel noticed that Sirius, which is one of the nearest stars, pursues a rather erratic course through the heavens. It weaves back and forth against the background of more distant stars.

Bessel concluded that this erratic behavior must be caused by an orbiting companion star, but he still hadn't found it when he died in 1844. US astronomer Alvan Clark was the first to spot the companion star 18 years later.

With a magnitude of about eight, the Companion of Sirius is theoretically bright enough to be visible in binoculars, but it gets lost in the glare of Sirius itself, brightest of all the stars.

right main image
Heart of the Dumbbell
With a temperature of around 180,000°F (100,000°C), the white dwarf at the center of the Dumbbell Nebula is brilliant white.

right inset
Blue skies
This glorious 2003 Hubble Heritage release reveals glowing knots of gas inside the Dumbbell Nebula. They seem to be a common feature in planetary nebulas. Here, the blue color indicates concentrations of oxygen gas.

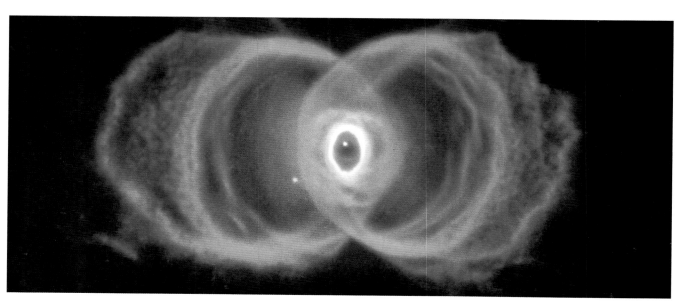

left
In the Hourglass
The white-dwarf star MyCn 18 lies at the heart of the Hourglass Nebula, some 8,000 light-years away. Note that it lies off-center— this suggests that it may have a faint companion.

the fifth-brightest star in the sky and one of the three bright stars that blaze overhead during the northern summer. The others are Deneb in the adjacent constellation Cygnus (the Swan) and Altair in Aquila (the Eagle). Together they form the famous Summer Triangle.

The other outstanding feature in Lyra is M57, the Ring Nebula. It is a planetary nebula—in fact, the definitive planetary nebula before the HST revealed the staggering beauty of so many others.

The Ring Nebula is easily located because it is sandwiched between Beta and Gamma, the second- and third-brightest stars in the constellation. It is not visible with binoculars, but even small telescopes reveal its "planetary" shape and smoke-ring appearance. Larger instruments are needed to show its central blue-white star, which is a white dwarf. It is one of the hotter types, with a surface temperature of around 180,000 degrees Fahrenheit (100,000°C).

Like other planetary nebulas, the Ring Nebula is growing as the shell of material ejected by the central star pushes further into space. The outer edge of the ring is well defined, which makes it possible to calculate its rate of expansion as close to 12 miles (20 km) per second. Working backwards, this gives an estimated date for the origin of the nebula of around 5,500 years ago.

Several other planetary nebulas have the appearance of cosmic smoke rings, although they are not as perfect as the Ring Nebula itself. One is the Helix Nebula in Aquarius (the Water-Bearer). It has two overlapping rings, which make it look like a spiral or helix.

hourglass shape. We see it as a ring because we happen to be looking at it end-on.

below left
Stellar smoke ring
With a magnitude of around 9, the Ring Nebula can be clearly viewed through small telescopes. Roughly 1 light-year across, it lies about 2,000 light-years away.

below right
The Helix
Overlapping rings of ejected material make the Helix Nebula NGC 7293 look rather like a coil spring. It is the largest planetary nebula.

New Stars

Sometimes a white dwarf forms part of a binary, or two-star system, along with a larger star. It then starts to attract gas — mostly hydrogen — from its companion. As more and more hydrogen rains down, a thick, dense layer builds up, and temperatures and pressures on the white dwarf begin to soar.

Eventually, conditions are extreme enough to trigger hydrogen fusion. A massive star-wide thermonuclear explosion blasts material into space and makes the star flare up to perhaps 100,000 times its original brightness in a few days. Over the following weeks it returns to its original state, until the same process begins all over again.

Such sudden flare-ups are called novas, which means "new." Early astronomers gave this name to what seemed to be new stars appearing in the heavens, when in fact they were existing stars that had been faint but suddenly became visible.

Supernova

Some "new stars" flare up to an even greater brilliance. They can increase millions of times in brightness and become as bright as a whole galaxy. We call these flare-up stars supernovas. On average, one supernova occurs in a galaxy about every 200 to 300 years.

The first recorded supernova seems to be the one that Chinese astronomers noted in 1054 in Taurus. We see its ghostly remains today as the Crab Nebula. The Danish astronomer Tycho Brahe spotted another supernova in our Galaxy in 1572, when it became as bright as Venus and was visible in daylight. It became known as Tycho's Star. In 1604, Tycho's protégé, the German astronomer Johannes Kepler, saw another supernova, which has been dubbed Kepler's Star.

That was the last time a supernova was seen in our Galaxy, although many have been spotted in others. Most notable was one in our galactic neighbor the Large Magellanic Cloud in 1987. Not yet operational at that time, the HST couldn't witness the event, but it has been used to study carefully the aftermath ever since.

We call the visible remains of a supernova explosion (such as the Crab Nebula) a supernova remnant (SNR). We find many other examples in the heavens, which often rival planetary nebulas in their delicate beauty.

Types I and II

There are two kinds of supernovas. A Type I supernova occurs in the same binary big-star/white-dwarf system that creates a nova. It is created when the white dwarf attracts so much matter from the larger star that it can no longer support itself. It collapses and destroys itself in a supernova explosion.

A Type II supernova occurs when a star with much greater mass than the Sun dies. Type II supernovas are more common. They release much more energy than Type Is, but not in visible light. Type I supernovas appear brighter.

left
The glorious crab
The Crab Nebula deserves its ranking as No. 1 in Charles Messier's catalog of nebulas and clusters. The remnant of a supernova seen nearly a thousand years ago, it is a writhing mass of expanding gas with a pulsar at its center.

right
Making waves
The supernova explosion that occurs when a massive star blasts itself apart sends powerful shock waves rippling through the surrounding space. These waves compress interstellar gas and make it glow, creating the luminous filaments we see in this image.

Eta Carinae

The HST has been used to examine closely one supermassive dying star: Eta Carinae, in the far southern constellation of Carina (the Keel).

Eta Carinae is one of the biggest, hottest and intrinsically brightest stars we know. Compared with the Sun, it is 100 times more massive, 150 times wider, and four million times more brilliant. It is also more than five times hotter than the Sun, with a surface temperature nudging 54,000 degrees Fahrenheit (30,000°C).

Eta Carinae has a visual magnitude of about seven, which puts it just beyond naked-eye visibility. But it is easily spotted with binoculars and a small telescope, and looks noticeably colored.

Astronomers know that a star with such staggering statistics must be inherently unstable, and so it has proved. Eta Carinae periodically flares up and then dims, fluctuating in brightness over irregular periods of time.

English Astronomer Royal Edmond Halley (of comet fame) watched the star brighten to the fourth magnitude in 1677. An even more spectacular flare-up began in 1835 and peaked in 1843, when Eta Carinae became the brightest star in the heavens after Sirius. Since then it has gradually dimmed.

Eta Carinae, then, is a variable star. It is called an eruptive variable because it varies in brightness when it undergoes massive eruptions from its atmosphere that blast matter into space. It flares up because the ejected matter is very hot and radiates abundant light.

Vast amounts of matter were ejected in Eta Carinae's massive eruption in the 1840s. Today we see the ejecta as an expanding red shell of nitrogen, oxygen and other gases, traveling at speeds up to two million miles per hour (3,000,000 km/h).

The violent eruption Eta Carinae suffered is reminiscent of a supernova explosion, but it was only a hint of what is to come. When this doomed star really goes supernova and blows itself to bits — which could happen any time — it will be a truly amazing sight.

right (main image)
Wide angle
A glorious stellar vista featuring the Eta Carinae Nebula is captured in this wide-angle photograph taken by a ground-based Schmidt telescope.

left
Flare-up
The HST has peered into the nebula to view the vast clouds of gas and dust ejected by the star Eta Carinae some 150 years ago. That was when it flared up to become one of the brightest stars in the heavens.

right inset
Doomed superstar
This later image shows spectacular detail of the massive clouds of ejected matter. It even glimpses the blue light coming from the intensely hot hidden star.

The Death of Massive Stars

Massive stars destroy themselves in a Type II supernova explosion. As was noted earlier, a star with about the same mass as the Sun rides the main sequence for billions of years, swells up into a red giant, puffs off its outer layers and then shrinks to become a tiny white dwarf.

But stars much more massive than the Sun live life in the fast lane. They consume their hydrogen fuel voraciously and spend only a few tens of millions of years on the main sequence. The biggest stars of all may spend less than a million years there.

As with a star like the Sun, a supermassive star stays on the main sequence while hydrogen fuses into helium in its core. When the hydrogen runs out, it starts swelling up, becoming cooler and redder. Because it is so massive, it swells up beyond the size of a red giant to become a supergiant, with hundreds of times the diameter of the Sun.

Inside a Supergiant

Like in a red giant, the core of a supergiant starts collapsing when it runs out of hydrogen. The collapse raises the temperature and initiates the fusion reactions that turn the helium into carbon and oxygen.

Once all the helium has been used up, the core begins to collapse again. Because the core in a supergiant is so massive, its collapse releases prodigious amounts of energy, pushing temperatures up to hundreds of millions of degrees. This is high enough to make the carbon and oxygen fuse to produce heavier elements, such as neon, magnesium, silicon and sulfur. These elements fuse in turn. The final product is iron, which is stable and will not fuse.

The rapidity with which these elements fuse is staggering. The carbon "burn" may take only about a thousand years, the oxygen burn just a year and the final silicon burn to form iron only a few days.

The Mother of all Explosions

Once the entire core of a supergiant has been converted into iron, its enormous mass causes it to collapse. The collapse is so rapid that the rest of the star crashes down on itself as well. The release of energy is catastrophic and blasts the star to smithereens. It becomes a supernova.

The temperatures and pressures produced in this cataclysmic event convert iron into a succession of heavier elements. In fact, all the elements heavier than iron that exist in nature were forged in supernovas. These mothers of all explosions blast newly made elements into interstellar space, where they make their way into the vast nebulas that spawn new stars.

How neat this is: the spectacular death of massive stars seeds the universe with the material to make new ones—celestial recycling.

Supernova SN 1987a

On February 23, 1987, astronomers spotted a bright "new star" in southern skies. It was a brilliant supernova, designated 1987a. Over the next weeks it nearly reached the second magnitude of brightness and was easily visible to the naked eye. It was the brightest supernova seen since Kepler's Star of 1604. But that star was not in our Galaxy; SN 1987a was in a neighboring galaxy, the Large Magellanic Cloud, some 160,000 light-years away.

The star that went supernova was identified as Sanduleak −69°202. It was a blue supergiant estimated to have had a mass 20 times that of the Sun. Shortly after the HST went into orbit in 1990, it spied a ring of matter that had been blasted into space by SN 1987a. It has been charting the incredible convolutions of this expanding ring ever since.

left
Supernova!
The International Ultraviolet Explorer (IUE) satellite recorded the initial outburst that accompanied the 1987a supernova in the Large Magellanic Cloud. The scale indicates level of brightness.

right
Glittering backdrop
This glittering starscape in the Large Magellanic Cloud provides a stunning backdrop for the remnants of supernova 1987a (center). Close by are many blue-white stars, shining brightly with youthful exuberance.

below
Bright ring
The HST first observed the site of the 1987a supernova in August 1990, when it spotted a glowing ring of material. This ring was probably in place already, as a result of previous stellar wind activity by the supergiant star. It brightened as a result of the intense radiation given out when the star exploded.

End of the Line

Exactly what happens to the collapsing core of a massive star in the aftermath of a supernova explosion depends on its mass. It will either become a neutron star or that most awesome of celestial objects — a black hole.

If the core has up to about three times the mass of the Sun, it becomes a neutron star. The force of collapse creates a new kind of matter. It forces the electrons in atoms into the nucleus, where they merge with the protons and turn into neutrons. The core becomes a mass of neutrons, tightly compressed together — a form of matter called baryon degenerate matter. (Neutrons belong to a class of subatomic particles called baryons.)

Such a neutron star is tiny, averaging only about 12 miles (20 km) across. Yet it contains three or more solar masses. So it has the most incredibly high density — millions of times greater even than a white dwarf.

Pulsating Signals

After a supernova explosion, the neutron star is left spinning furiously. Its powerful magnetic field spins as well, and funnels the star's energy into powerful beams of radiation that emanate from the two magnetic poles.

As the star rotates, these beams sweep around in space, rather like those from a lighthouse. If the beams are in our line of sight, we detect pulses of energy every time they sweep past. We call these pulsating bodies pulsars.

Little Green Men

Working at Britain's Cambridge Radio Astronomy Observatory, astronomer Jocelyn Bell-Burnell discovered the first pulsar in 1967 by its radio waves. Never before had such a regularly pulsating radio source been detected. Initially baffled over what it could be, the Cambridge astronomers dubbed it LGM (for Little Green Men), likening it to the kind of signals an alien civilization might beam into space.

Since then more than a thousand pulsars have been discovered. Most give off energy as radio waves, but some give off energy in other forms, including visible light, X-rays and gamma rays.

The first pulsar (PSR 1919+21) spins relatively slowly, sending out pulses every 1.3373 seconds. The fastest pulsar, PSR 1937+215, sends out pulses every 1.557 milliseconds (thousandths of a second). This means that it rotates 642 times every second. (Try to imagine something the size of New York City spinning around that fast!)

Black Holes

After a supernova, if the collapsing core has a mass greater than about three solar masses, it becomes a black hole. When such a massive core collapses, the powerful gravitational forces compress even neutrons, causing the body to go beyond the neutron-star stage. As the core gets ever smaller, the surface gravity gets ever larger, and the velocity needed for anything to escape from it escalates.

below
X marks the spot
This "first light" image from the X-ray astronomy satellite Rosat shows a powerful X-ray source in Cygnus that is probably a black hole in a binary system with a normal star.

left
Triple rings
In close up, two larger, fainter rings show up in the region of the 1987a supernova explosion, as well as the bright ring illuminated by its radiation. The rings look as though they intersect, but they probably don't. One of the larger rings appears to lie in front of, and the other behind, the bright central ring.

Eventually the surface gravity of the collapsed core becomes so high that not even light can escape from it. At this point it becomes a black hole and disappears from the observable universe. The circle that marks the edge of a black hole — the boundary between it and the visible universe — is called the event horizon. Its radius, called the Schwarzschild radius, depends on the mass of the collapsing body. Usually it measures just a few miles.

What lies within the event horizon is, of course, impossible to fathom. The theory is that the collapsing core continues to shrink until it becomes an infinitely small point of infinite density known as a singularity. The reality could be different — we will probably never know.

X Marks the Spot

Looking for black holes in the blackness of space is obviously impossible, but there are ways to detect them. When a black hole exists in a binary, two-star system, it attracts gas from the other star with its powerful gravity. As the two bodies orbit each other, the gas first swirls round the black hole, forming what is called an accretion disk.

The material in the disk swirls round very fast, and becomes searing hot because of friction between the particles. It gives off energy as X-rays before spiraling into the black hole. And we are able to detect these X-rays.

Cygnus X-1, a powerful X-ray source in the constellation Cygnus, is believed to come from such a binary black-hole system. It was the first object that astronomers suspected was a black hole, and they estimate that it contains the mass of nearly 20 Suns.

There are some black holes that have a mass that is millions of times the mass of the Sun. These supermassive black holes are found in the centers of galaxies, where they were produced by the collapse of enormous gas clouds, rather than by dying stars. They are the extremely powerful "engines" that are responsible for the extraordinary energy output of quasars and other active galaxies (see page 78).

above
In the Crab
The HST has homed in on the heart of the Crab Nebula to reveal the pulsar remnant of the 1054 supernova explosion. The pulsar is the left star of the bright pair in the center. It measures only about 6 miles (10 km) across and spins around 30 times a second.

3 | Gregarious Galaxies

The HST reveals the beauty of these great star islands in space

above
Galactic silhouettes
By a chance alignment, the HST has captured a view of one spiral galaxy superimposed against another, though the two are far apart. We see the smaller one in the foreground face-on, while the other presents a more edge-on aspect.

inset left
Spiral with a bar
Many galaxies have a spiral shape, like our own Galaxy, the Milky Way. Some have a distinct bar across the central bulge and are known as barred spirals.

inset right
Extragalactic
The HST spies the great clouds of gas and dust we call nebulas, not only in our own Galaxy but in others as well. It is within darker, cooler nebulas in our own and other galaxies that stars are born.

The Milky Way

In whichever direction you look in the night sky, you see twinkling stars, scattered in the velvety blackness of space. So is this what the universe is like—stars scattered haphazardly in space? Indeed it is not. If you could take a ride on an interstellar spaceship, head for the stars and then keep going some, you would eventually leave the stars behind. Looking back, you would see that all the stars in the sky are grouped together, forming a great star island in a vast ocean of space. And looking in other directions, you would see other star islands, with empty space in between. We call these separate star systems galaxies. Edwin Hubble was the first to prove the existence of external galaxies beyond our own and study them in detail.

Our own galaxy is called the Milky Way, but we often also refer to it simply as "the Galaxy." Like most other galaxies, it is a collection of billions of stars, bound together by gravity. It has a spiral shape and rotates slowly. From a distance it would look like a spinning wheel-shaped firework. Many galaxies are similar to it, but others are different. The most extraordinary ones pump out unbelievable amounts of energy. They include quasars—some of the most amazing and remote objects in the universe.

In the study of galaxies, the HST reigns supreme. With its incomparable clarity of vision and sensitivity, it provides panoramic vistas of deep space showing galaxies by the hundreds. It can also zero in on individual star systems to tease out the most minute details.

Look at the heavens on a clear, moonless night, and you'll see a faint, misty band of light arching across the sky. In the Northern Hemisphere, it runs through the constellations Cassiopeia, Perseus, Cygnus and Aquila, and in the Southern Hemisphere, Crux, Centaurus, Scorpius and Sagittarius.

What is this tenuous misty band? The ancient Greeks said it was a stream of milk spurting from the breast of the goddess Hera, the long-suffering wife of Zeus, the philandering king of the gods. They called it Kiklos Galaxias, meaning the milky circle. We know it today as the Milky Way. It was the Italian Galileo who first discovered the nature of the Milky Way, when he turned his telescope on the heavens in the winter of 1609-10. He saw that it was made up of countless numbers of faint stars, seemingly packed tightly together. If you look at the Milky Way through powerful binoculars, you'll see that he was right.

Later the following century, astronomers such as William Herschel began to realize what the Milky Way really is. It is a view from inside a layer of stars that form our star system—our Galaxy. By counting stars in regions of the sky on either side of the Milky Way, Herschel concluded that the Galaxy was lens-shaped: thickest at the center and thinnest at the edge. He was not far wrong.

Our Galactic Home

Our Galaxy, also called appropriately the Milky Way, is shaped somewhat like a disk with a huge bulge in the middle. Long, curved arms spiral out from the bulge, or nucleus. Rather inelegantly, the shape of the Galaxy has been likened to that of two fried eggs, stuck back to back!

The size of the Galaxy is astonishing. It measures 100,000 light-years from edge to edge. The nucleus is about 6,000 light-years thick, while the disk averages only about a third of this.

In all, the Galaxy contains as many as 200 billion stars. The Sun is located on one of the spiral arms, about 25,000 light-years from the center. The center lies in the direction of the constellation Sagittarius. And this is where the Milky Way appears brightest in the sky.

right
Face-on
This fine face-on spiral galaxy is M83. Relatively close at 15 million light-years, it has about the same mass and brightness as our own Galaxy. It is noted for its supernovas, which seem to occur more frequenty than in most other galaxies.

right
Night sky view
Small telescopes show the Milky Way well, revealing a dazzling panorama of close-packed stars, bright clusters and colorful nebulas. Many dark dust lanes are evident, particularly in the far southern constellations.

In the Center

The central bulge of the Galaxy contains mostly old red and yellow stars. There is comparatively little interstellar matter, and star formation is limited.

Until recently what lay in the interior of the bulge was a mystery, because thick clouds of gas and dust obscure our view. Radio and infrared studies of the region have now revealed a host of fascinating features.

At the exact center is a powerful radio source known as Sagittarius A. It appears to mark the site of a black hole with the mass of 2.5 million Suns. Astronomers believe that supermassive black holes like this might lurk at the center of most galaxies. Magnetism is also a dominant force in the galactic center, creating strange structures such as the Arc.

The Spiral Arms

Using radio telescopes to penetrate the obscuring clouds in the Milky Way, astronomers have created a detailed map of the Galaxy. These radio surveys have revealed the curved, spiral "arms" that make up the disk. Young, bright stars and bright and dark nebulas make up the shape of the arms.

There are two main spiral arms, Sagittarius and Perseus, as well as pieces of others. Our own Sun is on the Orion Arm, which forms a kind of bridge between the Sagittarius and Perseus Arms. Riding the arm in our neighborhood are stars of some of the most familiar constellations, such as Orion, Taurus and Cygnus. Clusters such as the Hyades and Pleiades lie relatively close, as do some of our most familiar nebulas, such as the Orion, Helix, Dumbbell and North America.

Farther away from the galactic center, the Perseus Arm is quite fragmented. It contains several well-known supernova remnants, including the Crab Nebula and Cassiopeia A. The Rosette Nebula in Monoceros and the double star cluster in Perseus are among other delights.

far left
On the radio
This radio image of the center of the Galaxy shows the curved feature called the Arc, which seems to be part of a ring of highly magnetized gas (the radio lobe). The white point is the radio source Sagittarius A.

left
On the Orion Arm
The Dumbbell planetary nebula is one of the many familiar night-sky objects in our local celestial neighborhood. It lies about 1,000 light-years away. Its dumbbell shape is more evident in less powerful telescopes than here.

above
Quintuplets
Near the center of the Galaxy, the HST has spotted a group of giant stars, called the Quintuplet Cluster. Ten times larger than typical clusters in the spiral arms, it is the home of the brightest star we know in the Galaxy, the Pistol Star (see page 177).

But it is on the Sagittarius Arm, closer in to the galactic center than the Orion Arm, that we find some of the most spectacular night-sky objects. Great billowing clouds of interstellar matter mark regions of intensive star birth in the part of the arms closest to us. They include the Eagle, Omega, Trifid and Lagoon Nebulas. The unstable giant star Eta Carinae threatens to go supernova, while the sparkling stars in Crux's Jewel Box cluster could not be more appropriately named.

On the Outside

Most of the matter in the Galaxy resides in the stars in the bulge and spiral arms, and in the interstellar medium — the gas and dust between the stars. But a few hundred tightly-knit globular star clusters circle outside the bulge and disk, and there are also large amounts of invisible dark matter within a great spherical halo that envelops the entire Galaxy.

Classifying the Galaxies

Our home Galaxy—a rotating spiral of hundreds of billions of stars that measures thousands of light-years across—is one of billions of such galaxies in the universe. They are known as spirals. Other galaxies consist only of a spherical or elliptical bulge of stars and lack the curved arms of spirals. They are known as ellipticals.

Spirals and ellipticals are termed regular galaxies because of their distinctive shape. But some galaxies have no particular shape and are termed irregulars. Edwin Hubble devised the method of classifying regular galaxies according to their shape in his so-called tuning fork diagram.

Elliptical Galaxies

More than half of all galaxies are ellipticals. They consist mainly of old stars and contain little free gas or dust, and so little or no star formation takes place within them. Elliptical galaxies vary widely in size and mass. The largest and some of the smallest galaxies are ellipticals. The giant elliptical galaxies at the heart of galaxy clusters measure many hundreds of thousands of light-years across.

At the other end of the scale are dwarf elliptical galaxies just a few hundred light-years across. Many of our galactic neighbors are dwarf ellipticals, some containing only a few hundred thousand stars, rather than billions.

In the Hubble classification, ellipticals are denoted E, followed by a number from one to seven describing how round or oval they are. E0 is a near-spherical galaxy and E7 the most flattened oval.

Spiral Galaxies

Spirals have essentially the same structure and composition as the Milky Way, with a flattened bulge at the center, and stars and dust carried on arms spiraling out. The bulge contains mainly older stars, with little gas and dust. It is orbited by a few hundred globular clusters. The arms, by contrast, are populated mainly by young stars, and star formation is rampant among the turbulent gas clouds of the interstellar medium.

left
Hubble's tuning fork
Edwin Hubble's system of classifying regular galaxies into ellipticals, spirals and barred spirals.

above
Dusty spiral
A beautiful spiral galaxy (NGC 4414) in Coma Berenices, which has copious amounts of dust in its spiral arms. As is usual in spirals, the central bulge has older, yellower stars, while the arms have many younger, bluer ones.

NICMOS

above
Deep south
Different HST views of
NGC 1365, in visible
light (left) and infrared
(right). NGC 1365 is one
of the brightest barred
spiral galaxies we know.
It is part of the Fornax
cluster of galaxies in
the far southern
constellation Fornax.

right
Seeing double
The HST reveals that
the cluster NGC 1850,
in the Large Magellanic
Cloud, is a double. The
dominant cluster is
made up of yellowish
stars around 40 million
years old. The scattered
white-hot stars in the
other cluster are ten
times younger.

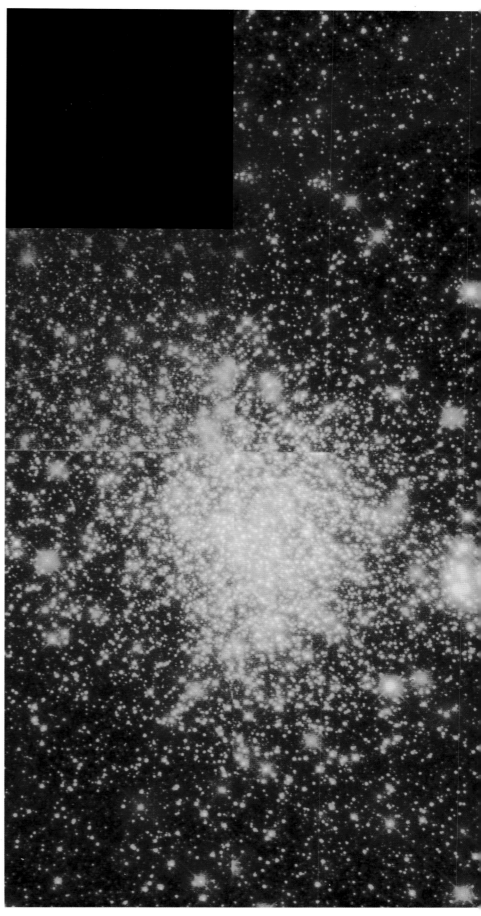

Spirals do not vary as much in size as ellipticals. The Milky
Way seems to be of about average size. In our galactic
neighborhood, the galaxy in Triangulum is less than half its size,
while the Andromeda Galaxy (see page 102) is half as big again.

Perhaps as many as half of all spirals have a distinct bar of
stars running through the nucleus. They are known as barred
spirals, and the spiral arms curve out from the ends of the bar.

In the Hubble classification, spirals are denoted S and
barred spirals SB. The letter a, b or c follows, indicating the
openness of the spiral arms: an Sa galaxy has the most closed
arms, an Sc the most open. Our own Galaxy is traditionally
classified as an Sb, since it has moderately open arms, though
with evidence of a slight bar through the nucleus, it is
sometimes classed as an SBb.

Irregular Galaxies
Irregulars, with no particular shape, are not all that common.
In general, they are comparatively small and contain relatively
few stars. Yet they are rich in clouds of gas and dust, in which
stars are being born. The two Magellanic Clouds visible to the
naked eye in far southern skies are irregulars (see page 76).

left
Hyperactive
The otherwise undistinguished southern constellation of Circinus hosts this unusual looking galaxy. It is a kind of active galaxy known as a Seyfert, noted for its particularly bright center, which seems to be ejecting gas into the surrounding space.

below
Starburst ring
Quite close, at a distance of 30 million light-years, this magnificent galaxy is NGC 1512. It is distinguished by a remarkable outer ring that is a cauldron of star formation. It is a barred-spiral galaxy, although the bar isn't actually visible in this image.

above
Edge-on
If we could view our own Galaxy edge-on from far out in space, it would look like this. This galaxy is NGC 4013 in Ursa Major. The dark band that cuts it in two is thick dust blotting out the light from background stars.

following pages
Galactic fireworks
Incandescent gas clouds and hot star clusters show that vigorous star formation is taking place in this nearby galaxy (NGC 4214). The central cluster is made up of hundreds of blue-white stars, each over 10,000 times brighter than the Sun.

Magellan's Clouds

The early sixteenth century was an exciting time to be alive if you were of an adventurous dispositionan. Italian-born Christopher Columbus had recently sailed from Spain across the Atlantic and discovered a New World, in 1492. Vasco da Gama had sailed from Portugal round the Cape of Good Hope to India , in 1498. This voyage fired the imagination of his young fellow countryman Ferdinand Magellan, then just 18 years old.

Twenty-one years later, now an experienced seaman and navigator, Magellan commanded an expedition that would become the first to circumnavigate the globe. Alas, he never completed the voyage, because he was killed in a skirmish with tribesmen in the Philippines in 1521.

Navigating for months under southern skies, Magellan would have become familiar with the dazzling southern constellations, particularly Crux (the Southern Cross). He would also have noticed the two misty patches visible nearby, and these were since named after him. They are the Large and Small Clouds of Magellan, or Magellanic Clouds.

Also named Nebecula Major and Minor, the Magellanic Clouds look like nebulas, but aren't. They are neighboring galaxies—two of only three galaxies we can see with the naked eye. The other is Andromeda (see page 102).

The Large Magellanic Cloud (LMC) is closest to us, at a distance of about 160,000 light-years. It contains much the same mix of stars and gas as our own Galaxy and has some structure. But at only about 30,000 light-years across, it is not big enough to develop into a spiral galaxy such as our own. One striking feature of the LMC is the Tarantula Nebula, named for its spidery appearance. It is one of the biggest and brightest nebulas we know and is easily visible through binoculars. The Small Magellanic Cloud (SMC) is only two-thirds as big across as the LMC and is about 30,000 light-years farther away.

The Magellanic Clouds are not only close neighbors in space, they are actually companions—satellites—of our own Galaxy. They circle our Galaxy once every 1.5 billion years or so, traveling in an elliptical orbit. Every time they make their closest approach, our Galaxy's powerful gravity attracts some of their stars and gas. As a result, the SMC is already showing signs of breaking apart. In time, both galaxies will be absorbed by our own.

right inset
Through the telescope
Telescopes show rather more structure in the LMC than in most other irregular galaxies. There is a distinct broad bar of stars in the center, measuring about 10,000 light-years long.

below
The Tarantula
The heart of the Tarantula Nebula, the most conspicuous part of the Large Magellanic Cloud (LMC) in visible light and infrared (insets). Also known as 30 Doradus, this nebula is a hotbed of star formation.

Active Galaxies

Galaxies give off energy as light, heat and other invisible radiation, such as X-rays and radio waves. Most of them give out the energy you would expect from a collection of billions of stars. But just a few — about one in 10 — give off exceptional energy, maybe millions of times greater than you might expect. They seem to pump out this energy from a tiny region at their center not much bigger than the solar system. Astronomers call these galactic mavericks active galaxies.

In 1943, US astronomer Carl Seyfert noted that some spiral galaxies have exceptionally bright centers. We now know that these Seyfert galaxies are one kind of active galaxy. Since then, several other kinds of active galaxies have been discovered, including radio galaxies, quasars and blazars.

Seyfert Galaxies

These active galaxies are among the most obvious because they mainly emit their exceptional energy at visible wavelengths. Analysis of their spectrum indicates that they contain clouds of hydrogen gas swirling at very high speeds around the galactic center.

Radio Galaxies

At radio wavelengths, these active galaxies appear among the biggest objects in the sky. The radiation seems to come not from the galaxy itself but from regions — radio lobes — on

above
Swan song
The second most powerful radio source in the heavens is located in Cygnus (the Swan). Called Cygnus A, it is an active galaxy that pumps out a million times more energy at radio wavelengths than our own Galaxy. The energy is broadcast mainly from two lobes, each 200,000 light-years from the galaxy's center.

below
Along the jets
Combined HST and radio images of the radio galaxy 3C-368 show a string of bright knots that may be stars or pockets of dust. This suggests that the jets streaming from black holes at the center of active galaxies might trigger star formation along their path.

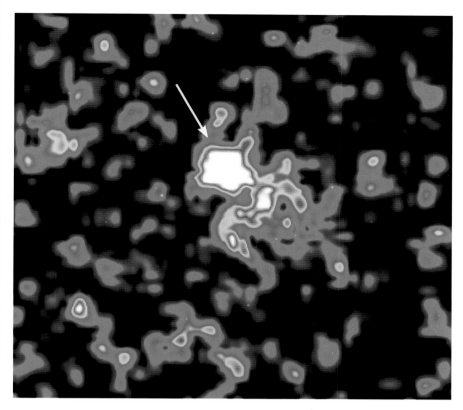

either side and very far out. Although the galaxy may measure only 100,000 light-years across, from lobe to lobe the radio source may span millions of light-years.

Quasars

In the third Cambridge catalogue of radio sources is one in the constellation Virgo, numbered 273, so it is identified as 3C-273. In 1963, when the Moon passed in front of this source and blotted out its signals, astronomers identified it with a star and took its spectrum.

When US astronomer Maarten Schmidt checked the spectrum, he found that it was like that of no other star he had ever seen. This "star" had a huge red shift of spectral lines that placed it at a distance of more than two billion light-years!

3C-273 was clearly no ordinary star, but quite a different body altogether. To be visible at such a distance, it had to be hundreds of times brighter than ordinary galaxies like the Milky Way. Because it appeared star-like, this newly identified celestial body was named a quasi-stellar radio source, or quasar.

Since that time, hundreds of quasars have been identified, giving off energy at X-ray and infrared wavelengths as well as light and radio waves. They are all remote, and some are the most distant objects we know in the universe. The quasar PKS 2000-330, for example, is at a distance of 13 billion light-years. This means that we are seeing it as it was close to the time when we believe the universe was born.

above
Gamma-ray burster
The HST spots the optical counterpart of an intense gamma-ray burst (GRB 970228) in the outer reaches of a remote galaxy. Astronomers reckon that the collision between two neutron stars might have triggered the violent release of energy.

left
Quasar host galaxies
The HST has targeted many quasars, revealing details of the host galaxies. Some galaxies seem remarkably undisturbed by quasar activity. In others, the quasars seem to be fuelled by the debris from collisions between galaxies.

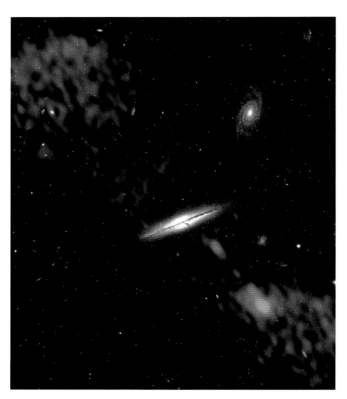

right
Spiral jet
The HST's Advanced Camera for Surveys (ACS) has spotted a giant radio jet coming from a spiral galaxy (0313-192). Here, the ACS image is shown superimposed on a radio image from the Very Large Array radio telescope. It is the first time that a radio jet has been spotted coming from a spiral galaxy.

below
Blowing bubbles
A black hole lurks at the center of NGC 4438, a member of the Virgo cluster of galaxies. Here, the HST has spied glowing bubbles of hot gas rising from the accretion disk surrounding the black hole.

Varying in Brightness

Quasars appear to be a very compact source of energy, much smaller than ordinary galaxies. But just how big are they? We can find out because of how their brightness varies. The brightness of a quasar can fluctuate on a timescale of a day or less.

But it is a fact that the shortest time in which an object can vary in brightness is the time it takes for light to travel across it. This means that a quasar that varies in brightness in a day must be only about a light-day across. This makes it comparable in size to our solar system.

Blazars

A blazar is another kind of active galaxy that varies widely at optical wavelengths. The name comes from a contraction of the terms BL Lacertae and quasar. BL Lacertae is an object first classified as a variable star but later found to be a strong radio source. It is notable because it does not show any lines in its spectrum.

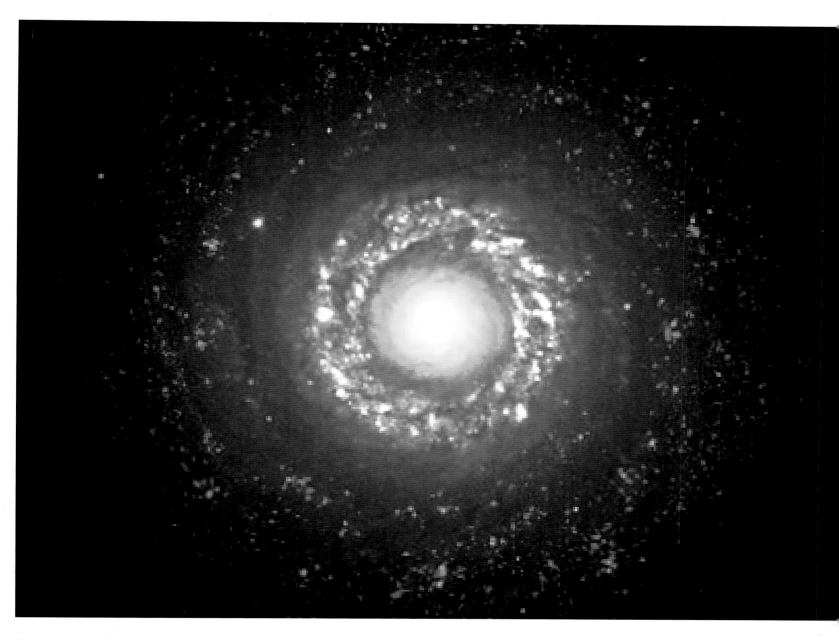

The Energy Machine

Clearly, for active galaxies to be able to pump out such prodigious energy, they must have a powerful "engine." Astronomers reckon that the only engine powerful enough is a supermassive black hole.

Black holes are so-called because they have such powerful gravity that nothing can escape from them—not even light. Relatively small black holes are created when big stars die (see page 62), but supermassive black holes are formed by the collapse of matter in the center of youthful galaxies.

Black holes produce energy when matter is sucked into them. But matter does not simply travel straight down. Because of a galaxy's rotation, matter is strung out into an accretion disk. Surrounding the disk is a donut-shaped torus (ring) of cooler gas and dust.

The accretion disk rotates rapidly and gets so hot that it emits X-rays and other forms of electromagnetic radiation. The radiation can't escape through the disk, and is instead beamed along the rotating axis. Subatomic particles produced in the whirling disk also pour out along the axis, forming high-speed jets. The jets give off radio waves as they encounter particles in the surrounding space.

Point of View

Astronomers believe that the various kinds of active galaxies are different views of the same black-hole structure. When our line of sight is along the plane of the disk, the cool torus obscures the bright center. But we can detect the radio-emitting regions on each side, created by the jets smashing into the intergalactic medium, and so we see a radio galaxy. When the disk is angled more toward us, we see a quasar or a Seyfert galaxy. And if the disk appears face-on, with a jet pointing directly at us, we see a blazar.

above
Sensational Seyfert
A stunning face-on view of the Seyfert galaxy NGC 7742. A massive black hole lurks in the very bright center of this active galaxy.

Centaurus A

With a visual magnitude of 7, Galaxy NGC 5128 is one of the brightest galaxies in the heavens, only just beyond naked-eye visibility. It is not included in Charles Messier's catalogue of clusters and nebulas and, therefore, does not have an M number.

This was not an oversight on Messier's part, but a reflection on the galaxy's location deep in southern skies and beyond Messier's reach. It lies in the constellation Centaurus (the Centaur) just north of Crux (the Southern Cross).

English astronomer John Herschel (son of William, who discovered Uranus) was one of the first European astronomers to study the southern skies, from South Africa, in the 1830s. He declared the galaxy to be "a most wonderful object, cut asunder by a broad, obscure band."

And NGC 5128 is a strange, beautiful object when viewed through a telescope. It is spherical in shape and is bisected by dark dust lanes. It lies about 15 million light-years away. In 1949, radio astronomers in Australia found a powerful radio source in Centaurus, which they called Centaurus A. They soon identified it with NGC 5128.

Astronomers speculated on the nature of this peculiar bisected galaxy: Could it be a spiral galaxy in the making, and the dark lanes part of a dusty disk that will eventually form spiral arms? Or could its appearance result from a collision between two galaxies — an elliptical and a spiral? Some astronomers favored the latter, reckoning that the radio emission was generated by the collision.

We now know that NGC 5128, usually called Centaurus A, is an active galaxy. It is the nearest active galaxy to us and the third most powerful radio source in the heavens (after Cassiopeia A and Cygnus A). It has been intensively studied by the most powerful radio telescopes and at other wavelengths by the HST, which peers deep into its mysterious dark dust lanes.

above
On the radio
This radio image of Centaurus A shows the typical wide-spaced lobes of radio galaxies. They extend about 20,000 light-years on either side of the galaxy's visible center.

below
Peering into the dust
In the HST image at left, clusters of hot blue stars line the edges of the dust lane, where star formation is most intense. An infrared view (right) shows a swirling mass of hot gas, caught up in the gravitational whirlpool of a massive black hole.

right
The bright center
In visible light, Centaurus A is one of the most intriguing and unmistakable objects in southern skies. The picture shows the bright central region of the galaxy, measuring about 30,000 light-years across, only about a fifth of its true extent.

Galaxies in Collision

The strange appearance of Centaurus A is not the result of an intergalactic collision. But collisions between galaxies do occur, and what celestial pyrotechnics they trigger.

Imagine two galaxies, each containing hundreds of billions of stars, closing on each one another at speeds approaching a million miles per hour. They are not solid objects, of course, and don't just bump and bounce apart. Each galaxy's stars begin to infiltrate the other's. The stars don't collide physically, because they are too far apart. Rather they run the gauntlet of a gravitational tug-of-war, being tugged one way by one galaxy, another way by the other. The ultimate outcome of the collision — which galaxy is the victor and which is vanquished — depends on the nature and relative mass of the two galaxies.

Stellar Streamers

If two spiral galaxies of similar mass collide, their stars are pulled from their original orbits, and the galaxies start to lose their spiral structure. Some of the stars gravitate toward the center of mass of the merging star systems. Others are flung out into intergalactic space in long streamers.

A billion years ago a high-speed collision between NGC 4038 and NGC 4039 produced a pair of long, curving streamers of stars. These give the merged galaxies the appearance of an insect's head and antennae, so the interacting pair are called the Antennae.

If the colliding galaxies are unequal in size, several things might happen. As the galaxies draw together, the smaller one could be pulled apart and merge imperceptibly with the larger one. This is happening with our own Galaxy, which is pulling apart its companions, the Magellanic Clouds (see page 76).

Or the small galaxy can also pass right through the large one, destroying its spiral structure and triggering a new round of star formation. This has happened spectacularly in the aptly named Cartwheel Galaxy, with vigorous star formation

above
The Mice
With their long tails of stars and gas, this interacting pair of galaxies (NGC 4676) is known as the Mice. They are located in the Coma cluster of galaxies, about 300 light-years away. Eventually they will merge into a single galaxy.

below
Cosmic ripples
The Cartwheel Galaxy
in Sculptor. This was
once an ordinary spiral
galaxy. Then came
catastrophe — another
galaxy crashed into it,
sending shock waves
rippling out into space,
pushing gas and dust
before it. Either of the
two nearby galaxies
might have been the
interloper since both
show signs of having
been disrupted.

following pages
The Tadpole
A stunning view of the
Tadpole Galaxy, which
has a long tail of stars
and gas streaming
from it. The tail has
been created during a
collision with a smaller
galaxy, visible in the
upper part of the
large spiral.

occurring at the "rim" of the wheel. Astronomers reckon that
the event happened around 500 billion years ago.

When the gas clouds in colliding galaxies crash together,
or are compressed by gravitational forces, they collapse. This
results in a sudden burst of star formation, an event called a
starburst. Starbursts also occur in galaxies that are being
distorted by larger ones nearby. A good example is M82, which
is being disturbed by the larger M81 nearby.

Mergers and Takeovers
Generally, when small and large galaxies interact, the larger
one swallows up the smaller and gets even bigger. Over time,
astronomers reckon that most galaxies will merge with their
neighbors. It seems likely that our own Galaxy,
for example, will merge with the Andromeda Galaxy in
about five billion years. So, gradually, the nature of the
universe will alter as it becomes populated by fewer but
larger galaxies.

In the Whirlpool

William Parsons, the third Earl of Rosse, was an Irish landowner, whose home was at Birr Castle, Parsonstown, not far from Athlone in central Ireland. Upon graduating from Oxford in 1822, he first dabbled in politics, but abandoned that for astronomy in 1834.

Not content to buy a telescope, Lord Rosse decided to build his own. By 1838, he had built a reflector with a 36-inch (91-cm) mirror. It was an excellent instrument which encouraged him to attempt one twice as large. No one had ever tried to build one this size before. In 1845, he completed the 72-inch (1.8-m) reflector. It was a monster. The mirror was housed in an iron tube nearly 60 feet (18 m) long and 8 feet (2.4 m) across. It was mounted between two massive stone walls and was aptly named the Leviathan of Parsonstown.

With his new telescope, Lord Rosse decided to start observing the star clusters and nebulas that Charles Messier had listed in his catalogue. When he turned the Leviathan on the nebula with the Messier number 51, he was astonished to find that it was shaped like a pinwheel. It had "spiral convolutions," he said. He was the first to discover the spiral nature of the nebulas that would in the 1920s be recognized as external galaxies.

M51 lies about 20 million light-years away in the constellation Canes Venatici (the Hunting Dogs). It is easy to find with powerful binoculars or a small telescope, because it lies just south of Alkaid, the first star in the handle of the Big Dipper. Its nickname, the Whirlpool Galaxy, is apt because we see it face-on, with its spiral arms beautifully revealed.

M51 is actually not one galaxy, but two. The main spiral is NGC 5194, and one of its arms is tenuously linked with a smaller galaxy, NGC 5195. The present structure has resulted from a glancing collision between the two galaxies that probably took place around 300 million years ago. The smaller galaxy was by then an ordinary spiral, or more likely a barred spiral. As it slammed into the outer regions of the large spiral, NGC 5195 was torn apart. Most of its stars were strung out to form a bridge with the other galaxy.

For its size, M51 is particularly bright, with an abundance of young, hot blue stars. Though only about two-thirds as big across as the Milky Way, it is three to four times brighter. Powerful bursts of radio waves come from the centers of both galaxies. The exceptional activity taking place in them almost certainly stems from the collision, as usually happens when galaxies interact.

right
Heart of the Whirlpool
The close encounter of small companion galaxy NGC 5195 (off the upper edge of the picture) is triggering intensive star formation in M51. The clusters of young stars are highlighted in red, the characteristic radiation from glowing hydrogen gas.

below left
Bridging the gap
This wider-angle ground telescope image shows well the Whirlpool's widespread spiral arms and the bridge of gas between it and its companion.

below right
In false color
This image demonstrates the combined results of imaging the Whirlpool Galaxy at different optical and radio wavelengths, shown here in the different colors.

4 | The Expansive Universe

The universe might continue to expand forever

above
Cosmic magnifier
This cluster of galaxies, known as Abell 2218, resides in the northern constellation Draco. Its combined gravity is huge and bends light like a lens. This gravitational lens effect brings more distant galaxies into view as a series of arcs.

inset left
Ancient globulars
Here the HST has imaged the center of a globular cluster (M22) in Sagittarius, where the stars are 100,000 times denser than in the Sun's neighborhood. Most of the stars in the cluster are ancient — they are probably more than 10 billion years old.

inset right
Standard candle
With its phenomenally sharp vision, the HST helps pinpoint "standard candles" in remote galaxies. These objects help astronomers make more accurate estimates of the scale of the universe.

After the Big Bang

Everything that exists—our bodies, the air we breathe, the Sun, the stars, the galaxies and space itself—makes up the universe. From year to year we see our local universe, the solar system, change markedly. Changes in the stellar heavens and in the universe at large are much more subtle, but they do occur.

The universe is changing and evolving. But where did it come from? How has it evolved? Exactly how is it made up? And what will happen to it in the future? The astronomers who seek to answer such fundamental questions are known as cosmologists. Cosmology is the study of the origins and evolution of the universe.

Irishman James Ussher, who became archbishop of Armagh in 1625, was no cosmologist, but using biblical references he calculated that the creation of the Earth and the heavens took place on an October morning in the year 4004 BC. He was far off—by a factor of nearly a million. Studies of radioactive rocks suggest that the Earth, along with the rest of the solar system, was born about 4,600 million years ago.

The universe itself is much older. Astronomers believe that it came into being between about 12 and 15 billion years ago. How do they know this? Well, they find that the galaxies are all rushing away from us and from one another. The universe seems to be flying apart, expanding, as though from a massive explosion long ago.

And astronomers reckon that a kind of explosion really did take place—they call it the Big Bang. By measuring the rate at which the universe is expanding now, and calculating backwards, they have determined when the explosion happened—when the universe was created. It was between 12 and 15 billion years ago.

What happened before the Big Bang? This question is impossible to answer because when the Big Bang happened, time itself began. So there can be no "before."

Cosmologists think they know what happened after the Big Bang—except for the first 10^{-43} seconds. During this period, called Planck time, space and time itself, the fundamental forces of nature and the laws of physics were still forming. After this time, cosmologists can apply knowledge of fundamental forces and laws to describe what happened next.

The universe remained tiny until it was 10^{-35} seconds old. It was incredibly hot (around 10^{27} degrees) and full of energy. Gravity had already become a distinct force, and the first fundamental particles, including electrons, began to form. The universe then expanded suddenly, in an event called inflation. As the universe ballooned, its temperature dropped rapidly.

When the universe was one-millionth (10^{-6}) of a second old, protons and neutrons started to form. Protons are the nuclei (centers) of hydrogen atoms. By about three minutes, the temperature of the universe was only about one billion (10^9) degrees. Protons and neutrons began to stick together, forming the nuclei of deuterium (heavy hydrogen) and helium. At this time the proportions of hydrogen and helium in the universe were fixed.

For about the next 300,000 years, things quietened down somewhat. The universe continued expanding and cooling, but it stayed essentially the same—full of radiation and a mass of atomic nuclei in a sea of electrons. It was a "foggy" universe. Because of all the particles milling around, radiation couldn't travel far without being scattered, like light in a fog

About 300,000 years after the Big Bang, temperatures fell to about 5,000 degrees Fahrenheit (3,000°C). Electrons could now combine with nuclei to form the first atoms of hydrogen and helium. The "fog" cleared as particle numbers were drastically reduced, allowing radiation to pass relatively unhindered. The universe became transparent.

Cosmologists estimate that some time within the first one or two billion years, hydrogen and helium began to form clouds. These would later collapse and create the first stars and galaxies. The HST is now tracking galaxies over 10 billion light-years away, which means we are seeing them as they were 10 billion years ago, when they were still very young.

right (main image)
Hubble Deep Field
Remote galaxies that we see as they were more than 10 billion light-years ago. This landmark image, named the Hubble Deep Field, is the result of a 120-hour exposure of a small region of the heavens just to the north of the Big Dipper (Plow). We are looking at galaxies in their infancy, just a few billion years after the universe was born.

right insets
Building blocks
Peering deep into space in the constellation Hercules, the HST has imaged 11-billion-year-old objects that could be the building blocks of galaxies. These objects are clumps of a few billion young stars, averaging about 2,000 light-years across.

right
Fireworks finale
The very first stars that formed may have burst into the universe as brilliantly and spectacularly as a fireworks finale, an event depicted in this artist's impression. Future generations of space telescopes may well be able to capture this defining moment in the history of the universe.

The Scale of the Universe

How big is this universe of ours? Unimaginably vast. But this is a relatively recent realization. Early astronomers knew that the universe was bigger than the Earth (which they thought was its center), but they did not believe it was much bigger. Later, astronomers coming to terms with Copernicus's concept of a solar system, equated this solar system with the universe.

Starting in William Herschel's time, in the eighteenth century, people considered the Milky Way Galaxy to be the whole universe. In 1918, US astronomer Harlow Shapley calculated its size to be 300,000 light-years; it turns out this estimate was three times too big. But the true scale of the universe did not became apparent until five years later, when Edwin Hubble proved that spiral "nebulas" were actually separate and remote galaxies, millions of light-years away.

Since then, astronomers have progressively pushed back the boundaries of the universe, using ever-more-powerful telescopes and, more recently, space telescopes like the HST.

From Atoms to Superclusters

The tiniest objects that exist in the universe are particles within atoms—the basic constituents of ordinary matter. Atoms measure about 10^{-10} meters across, or 1/10,000,000,000th of a meter (atomic sizes are expressed as fractions of a meter, never in fractions of an inch). This means that it would take 10 billion atoms, side by side, to measure a meter. A hundred thousand times smaller still, at around 10^{-15} meters across, are the electrons that circle the atomic nucleus. They are among the smallest of all atomic particles, slightly bigger than particles known as quarks (10^{-16} meters).

We need to scale things up a billion times or so to bring us into the realm of our everyday life, when measurements are made in centimeters and meters (or inches, feet, yards, and so on). A few million times bigger is the Earth itself, at about 10^7 meters. And a billion times bigger still is the entire solar system, at about 10^{16} meters (10^{13} km).

When the numbers get this large, measurements expressed in meters or even kilometers become meaningless, so we switch to expressing distances in light-years: one light-year is about 10^{13} km. The solar system is thus about one light-year across.

Multiply the size of the solar system by 100,000, and we get the size of the next-biggest entity, the Milky Way Galaxy. Multiply it 50 times more, and we're up to the scale of the cluster of galaxies to which the Milky Way belongs. Twenty times more, and we're looking at a supercluster. Two-hundred times more, and we're nearing the edge of the observable universe, more than 10 billion light-years away.

This distance works out at 10^{26} meters. The scale of the universe—from the smallest object to the largest, from a quark to the observable edge of space—goes from 10^{-16} to 10^{26} meters. In other words, the universe is more than a million, million, million, million, million, million, million (10^{42}) times bigger than the smallest particle.

Cosmological Distancing

Proxima Centauri, the nearest star to the Sun, lies about 4.3 light-years away from Earth. Sirius, the brightest star in the heavens, lies 8.8 light-years away. The Andromeda Galaxy, the farthest object we can see with the naked eye, lies 2.5 million light-years away. Some galaxies lie more than 10 billion light-years away.

How do we know that these figures are accurate? How do we measure the distance to the stars and galaxies? To measure the distance to a few hundred of the nearest stars, astronomers use a variation of the technique surveyors use to measure distances on land.

In this process, called triangulation, a surveyor first measures out a baseline — a straight line along the ground — and then measures the angles from each end of the baseline to a distant point, for example, a church. Knowing the length of one side (the baseline) and two angles, the surveyor can use simple trigonometry to calculate the lengths of the other two sides of the triangle formed by the ends of the baseline and the church. This pinpoints the location of the church.

Using Parallax

For the astronomical version of triangulation, astronomers choose a much longer baseline: the diameter of the Earth's orbit around the Sun. They view a star from one side of Earth's orbit in January, for example, and then from the other side six months later. A nearby star seems to change position slightly against the background of distant stars during this time. This

right
M100
The remote spiral galaxy M100, one of thousands in a mammoth cluster of galaxies in Virgo. The HST can pinpoint individual stars in the spiral arms, in particular Cepheid variables, which can be used as standard candles.

below
Hipparchos
Named for the foremost astronomer of ancient Greece, Europe's Hipparchos satellite was launched in 1989. It spent over three years measuring the parallax of some 120,000 stars with unprecedented accuracy.

April 23

May 4

effect is called parallax. You can easily demonstrate parallax yourself. Hold up a finger in front of you and look at it first with one eye and then the other. You'll notice that your finger seems to shift position against the distant background, and the nearer your finger is to your face, the more it seems to shift.

From the parallax shift shown by a star, astronomers calculate its distance in the same way a surveyor would for the church. However, parallax measurement has severe limitations. Parallax shifts are tiny, even for the nearest stars, and for stars farther than about 100 light-years, they are almost impossible to detect. In the early 1990s, however, the European astronomy satellite Hipparchos managed to measure parallaxes of around 120,000 stars out to a distance of about 500 light-years.

Standard Candles

Parallax measurement is certainly not an option for measuring distances to other galaxies. To measure galactic distances, astronomers look for what are called standard candles. These are objects that are always of similar brightness or behave in a predictable way.

The classic standard candles are variable stars called Cepheids. Early last century, US astronomer Henrietta Leavitt discovered a fundamental law about Cepheids: that the period over which they vary in brightness is directly related to their true brightness (absolute magnitude or luminosity).

This relationship, known as the period-luminosity law, makes Cepheids standard candles. When astronomers measure the period of a Cepheid, they know what its true brightness is. They then compare this true brightness with the star's apparent brightness in the sky. Because light fades in proportion to increasing distance, it is then easy to calculate how far away the Cepheid is.

It was Edwin Hubble who first used a Cepheid as a standard candle to measure the distance to the Andromeda Galaxy, proving that it lay beyond our own Galaxy (see page 162). Today the HST is extending his work, pinpointing Cepheid standard candles in some of the most distant galaxies.

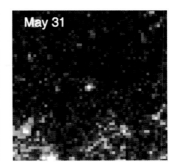

left
Cepheid variations
From HST observations of the light variations of a Cepheid variable in M100 over several months, astronomers find its period and hence its luminosity, or true brightness. Comparing this with its apparent brightness, they calculate that the Cepheid lies 51 million light-years away.

Other Standard Candles

Globular clusters are other objects that can be used as standard candles. These globe-shaped masses of stars orbit the centers of galaxies, and the brightest ones have about the same true brightness. By comparing this with their apparent brightness in the sky, astronomers can estimate their distance.

Supernovas are useful as standard candles for even more remote galaxies. These exploding stars can become for a while as brilliant as an entire galaxy and can therefore be spotted at enormous distances. The most useful kind of supernova is the Type I, caused by the explosion of a white dwarf star (see page 57), because they all reach close to the same level of brightness.

Red Shift

The most remote galaxies are too far away to detect any kind of standard candle. To determine how far away they are, astronomers revert to the invaluable astronomical technique of spectroscopy.

Just as astronomers can tell a lot about a star by examining the spectrum of its light (see page 32), they can also tell a lot about a galaxy from its spectrum. They see in the spectrum the characteristic dark absorption lines, which identify certain chemical elements. They can tell by the position of the lines in the spectrum whether the galaxy is moving toward or away from us.

below
Standard globulars
Globular clusters can make useful standard candles to measure distances to some galaxies. Here in the core of the giant elliptical galaxy NGC 1275 in Perseus, the HST resolves a number of globular clusters (blue). These are unusual because they consist of young stars, not old.

When a galaxy is moving toward us, its light waves bunch up and seem to have a shorter wavelength, which makes them appear bluer. And the galaxy shows a spectrum with lines shifted toward the blue end. This is called blue shift. When a galaxy is moving away, its light waves stretch out and seem to have a longer wavelength, which makes them appear redder. And the lines in the spectrum shift toward the red end. This is called red shift. The extent of the blue or red shift tells us how fast the galaxy is moving.

This bunching up and stretching out of waves from a moving source is called the Doppler effect. We experience this with sound waves when an ambulance — siren blaring — races first toward us and then away. The noise of the siren has a higher pitch (shorter wavelength) when the ambulance is approaching, because the sound waves are bunching up. The noise has a lower pitch (longer wavelength) when the ambulance is going away, because the sound waves are being stretched out.

Relating Red Shift

Astronomers find that almost all the galaxies have a red shift, indicating that they are rushing away from us and from one another. This is how Hubble and other astronomers in the early twentieth century realized that the universe must be expanding.

Hubble then made a discovery that is crucial to cosmological distance measurement. He found that the speed of recession of a galaxy (measured by its red shift) is directly related to its distance. This relationship is known as Hubble's Law. The rate at which galaxies speed up with increasing distance is called the Hubble Constant. Current calculations seem to show that a galaxy speeds up by around 50 miles (80 km) a second every megaparsec (about three million light-years).

By more accurately measuring the distance to remote galaxies, the HST has been helping refine the value of the Hubble Constant. The more accurate the value, the more accurately we can describe and comprehend the universe.

above and left
Ancient blast
Comparing the light output of galaxies in the Hubble Deep Field over an interval of two years, HST scientists discovered in one of them a supernova. The explosion had taken place 10 billion years ago. Supernovas prove useful in estimating distances to remote galaxies.

left
Spectral shifts
The shift of the dark lines in the spectra of stars and galaxies tells us if they are traveling toward us (blue shift) or away (red shift). Spectral shifts can also detect rotations. Here, the HST's spectrograph has scanned the rotating disk around a black hole in the galaxy M84. The colors show the abrupt shift in wavelengths from one side of the disk to the other.

Grouping Together

Galaxies are not distributed evenly throughout space; they gather together into small groups or larger clusters. It is the mutual gravity of galaxies that helps keep them together and overcomes, on a small scale, their tendency to fly apart as the universe expands.

In our corner of space, gravity draws together our Galaxy (the Milky Way) and about 30 other galaxies to form what is called the Local Group. The Milky Way is one of the two large spirals that dominates the Local Group. The other is the Andromeda Galaxy. There is just one other smaller spiral, M33. All the rest are small, dwarf elliptical or irregular galaxies. Altogether, the Local Group occupies a region of space around five million light-years across.

Satellite Galaxies

Within the Local Group, smaller galaxies congregate around the two large spirals. We call them satellite galaxies. The two Magellanic Clouds we can see with the naked eye in southern skies are satellites of the Milky Way. Our Galaxy also has many other satellites, all of them dwarfs and all too far away to see with the naked eye.

The Sagittarius dwarf galaxy, about half as wide across as the Large Magellanic Cloud (LMC), lies even closer to us, at a distance of around 78,000 light-years. This is so close that the gravity of the Milky Way is ripping it apart. Obscured by the dust in the plane of the Milky Way, this galaxy was not discovered until late last century.

Since the Sagittarius dwarf galaxy is nearly spherical, it is classified as elliptical. Many of the Milky Way's other satellites are dwarf ellipticals. They include the diminutive Draco and Carina systems, both around 250,000 light-years away and both only about 500 light-years across — only 1/200th the size of our own Galaxy. They may contain as few as 300,000 stars.

The Other Spirals

The Andromeda Galaxy is a spiral, just like the Milky Way but considerably larger. This magnificent galaxy has figured prominently in the history of astronomy and is profiled overleaf.

The Andromeda Galaxy also has satellites orbiting round it. The third spiral in the Local Group, M33, seems isolated, although many astronomers consider it to be a satellite too.

M33 is often called Triangulum because it lies within the tiny northern constellation of that name, whose three brightest stars form the shape of a triangle. M33 looks magnificent from Earth because we see it face-on and have an excellent view of its spiral arms. With a visual magnitude of just under six, M33 is right at the limit of naked-eye visibility. Observers with exceptional eyesight might spot it under very clear conditions. However, it is easy to see with binoculars, visible in the same field of view as the constellation's lead star, Alpha. One striking feature of the galaxy is a giant nebula of hydrogen gas, NGC 604, which is a vast star-forming region on one of M33's spiral arms.

above
Wide-open arms
The wide-open arms of M33 are well seen in telescopes. The galaxy lies 2.7 million light-years away, a little farther away than the Andromeda spiral. With a diameter of some 40 million light-years, it is less than half the size of our own Galaxy.

right
Star-birth spectacular
NGC 604, in the outer reaches of the Triangulum Galaxy (M33), is one of the largest star-forming regions we know. At the heart of this spectacular nebula are more than 100 hot newborn stars. The intense ultraviolet radiation they give out excites the surrounding gas, creating an object of breathtaking beauty.

The Great Spiral

The constellation Pegasus, the Flying Horse, is a prominent feature of fall skies in the Northern Hemisphere (and spring skies in the Southern). It is easy to recognize because it contains an almost perfect square, the Square of Pegasus.

The star that marks the top-left corner (bottom-right in the Southern Hemisphere) actually belongs to the linked constellation Andromeda. If you cast your eyes up and slightly to the left (down and to the right in the Southern Hemisphere), you will see a faint, misty patch. It looks like a nebula, but isn't — it's a neighboring galaxy.

The Andromeda Galaxy is nowhere near as close as our other galactic neighbors, the Magellanic Clouds. It lies around 2.5 million light-years away and is the most distant object that is visible to the naked eye.

The Arabian astronomer Al-Sufi mentioned the galaxy first in his *Book of the Fixed Stars* in the year AD 964. German astronomer Simon Marius spotted it again in a telescope in 1612, saying that it looked like "the light of a candle seen through horn." Charles Messier included it as number 31 in his list of star clusters and nebulas of 1781.

It seems astonishing that we can see the Andromeda Galaxy, M31, at a distance of millions of light-years away. This is because of its enormous size — 150,000 light-years across, which is one-and-a-half times as big as the Milky Way. It also contains twice as many stars — some 400 billion. It dominates the Local Group and could, in billions of years, consume most of the other galaxies, including the Milky Way.

M31 is a spiral galaxy, also called the Andromeda Nebula and the Great Spiral. Unfortunately, we don't have a very good view of its spiral arms because we see it from the side. M31 was one of the spiral nebulas investigated in the early twentieth century by Edwin Hubble and other astronomers. It was the first one he proved was extragalactic — beyond our own Galaxy (see page 162).

Later studies of the spiral arms and central bulge by Walter Baade (1944) led to his recognition of the two populations of stars. Population I stars are the relatively young stars of spiral arms, while Population II stars are the older stars of the galactic center.

The Andromeda Galaxy is like our own spiral in another way; it has two close companions, the galaxies M32 and NGC 205. They are both dwarf elliptical galaxies in orbit around the Great Spiral.

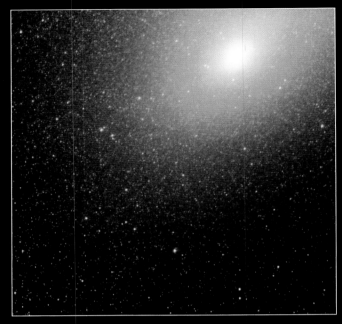

above
Stars with the blues
Deep inside the elliptical satellite galaxy M32, the HST has spotted a swarm of hot blue stars, seen here in ultraviolet light. Unusually, these stars seem to be old, at a late stage in their evolution when they are burning helium in their cores.

below
Andromeda's globulars
The HST views globular clusters orbiting M31's nucleus with almost the same clarity as ground-based telescopes view globulars in our own Galaxy. This Andromeda globular boasts plenty of red giants, as would be expected. The two other bright objects are nearby stars in our Galaxy.

right
The spiral giant
Our close galactic neighbor, the Great Spiral in Andromeda (M31), is the most famous of all the outer galaxies. It appears to have two major spiral arms. Seen here also are the two satellite galaxies — M32 close in and NGC 205 further out.

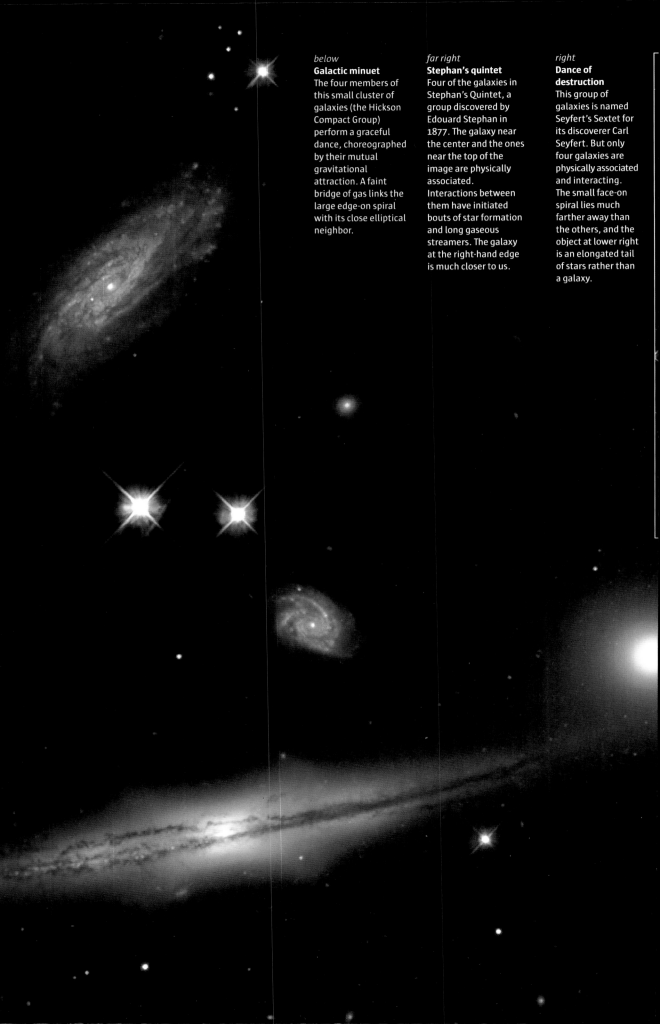

below

Galactic minuet

The four members of this small cluster of galaxies (the Hickson Compact Group) perform a graceful dance, choreographed by their mutual gravitational attraction. A faint bridge of gas links the large edge-on spiral with its close elliptical neighbor.

far right

Stephan's quintet

Four of the galaxies in Stephan's Quintet, a group discovered by Edouard Stephan in 1877. The galaxy near the center and the ones near the top of the image are physically associated. Interactions between them have initiated bouts of star formation and long gaseous streamers. The galaxy at the right-hand edge is much closer to us.

right

Dance of destruction

This group of galaxies is named Seyfert's Sextet for its discoverer Carl Seyfert. But only four galaxies are physically associated and interacting. The small face-on spiral lies much farther away than the others, and the object at lower right is an elongated tail of stars rather than a galaxy.

Clusters and Superclusters

The Local Group is a very small cluster of galaxies. But about 50 million light-years away in the constellation Virgo there is an enormous cluster. The Virgo cluster contains around 2,000 galaxies, spread out over a region of space some 10 million light-years across. A more distant cluster in Coma Berenices contains more than 3,000 galaxies in a region twice the size.

Clusters of galaxies vary not only in size but also in shape. The Local Group and the Virgo Cluster are irregular in shape, but the Coma Cluster is roughly spherical.

In general, irregular clusters are made up mainly of spiral galaxies, while elliptical galaxies dominate in spherical clusters. Irregular galaxies also seem to be much younger than spherical ones. Astronomers think that this probably reflects how galaxies evolve. Over time, the spirals in irregular clusters tend to collide and merge to form ellipticals, while the cluster itself assumes a more regular shape.

Galactic collisions account for the presence in some large clusters of giant elliptical galaxies. There are three at the center of the Virgo Cluster. Giant ellipticals are worthy of their name because they can measure as much as two million light-years across — about half the size of the entire Local Group.

Superclusters

It is gravity, of course, that holds the galaxies together in clusters. Gravity also binds clusters together to form mammoth superclusters. Our Local Group belongs to a supercluster centered on the big Virgo cluster. Called the Virgo, or Local Supercluster, it occupies a region of space about 100 light-years across.

Superclusters are the biggest structures in the universe. So far more than 50 have been discovered. They may take a variety of shapes, from flat sheets (called filaments) to curving streamers. The largest one we know is a filament called the Great Wall. It measures about 750 million by 250 million light-years.

Like stars and galaxies, superclusters are not distributed evenly throughout space. They seem to be wrapped around empty spherical regions called voids. The voids connect with one another, giving the universe a spongy structure.

This irregular scattering of galaxies and clusters could well reflect the original "lumpiness" of matter after the Big Bang, as inferred from measurements of cosmic background radiation by the COBE satellite (see page 156).

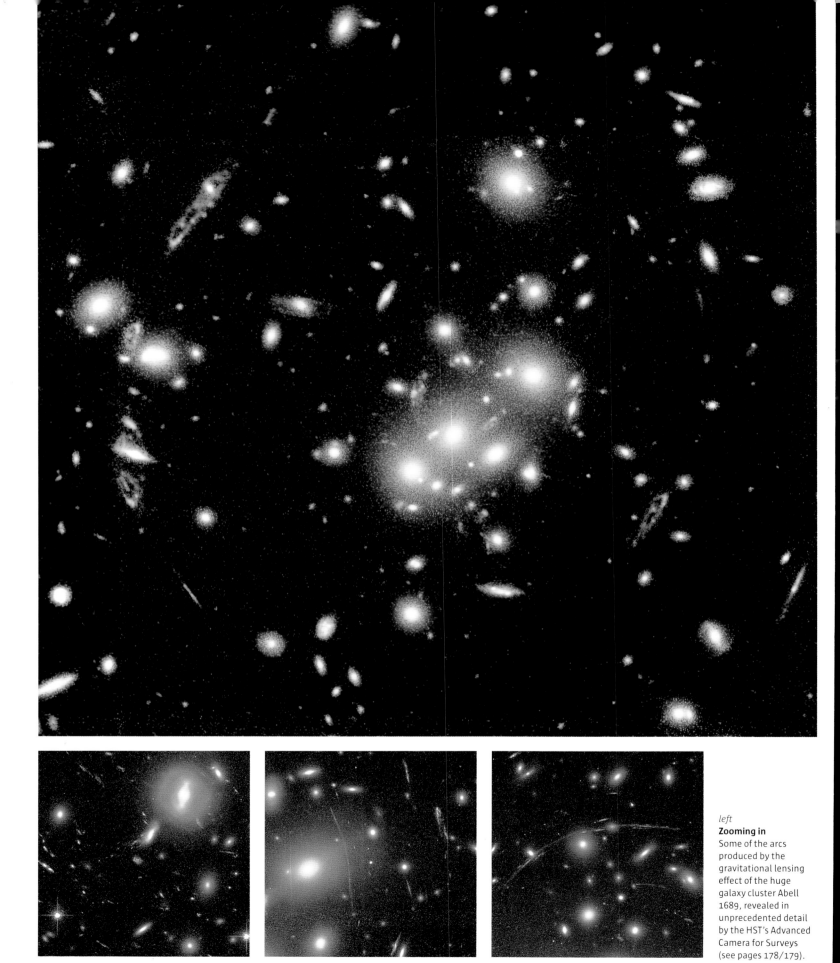

left
Zooming in
Some of the arcs produced by the gravitational lensing effect of the huge galaxy cluster Abell 1689, revealed in unprecedented detail by the HST's Advanced Camera for Surveys (see pages 178/179).

A Matter of Gravity

left
Copycats
A superb example of gravitational lensing, created by a cluster of elliptical and spiral galaxies known as 0024+1654. The cluster's lensing effect produces a pattern of arcs (blue). The similar shape of the arcs and their similar red shifts shown that they are views of the same object, a spiral galaxy about 10 billion light-years away.

Some of the most interesting HST images show a phenomenon known as gravitational lensing. Some pictures of distant clusters of galaxies reveal little curved arcs, which appear to be separate, distorted galaxies.

The red shifts of the light from these arcs place them far beyond the cluster itself, and the red shift of all the arcs is the same. It turns out that what we are seeing are repeated, distorted images of the same galaxy.

To find out why this is happening, we must refer back to Albert Einstein's theories of relativity. In his General Theory, published in 1916, he pointed out that gravity can interact with light and bend it. In our example, the cluster has such powerful gravity that it bends the light coming from the distant, hidden galaxy and brings multiple images of it into view. The cluster is acting like a lens—a gravitational lens.

Exactly how a distant galaxy appears as a result of gravitational lensing depends on how the galaxy and the cluster are aligned toward us. If the alignment were perfect, the arcs would merge into a complete, or "Einstein" ring.

The Riddle of Dark Matter

When astronomers calculate how strong the gravity of a cluster must be to produce a lensing effect, they come up against a problem. Without exception, the total mass of the galaxies visible in the cluster is not nearly large enough to produce the powerful gravity needed for lensing. In general, a cluster needs 90 percent more mass to produce this effect. Astronomers reckon that this mass must reside in matter that is invisible. They call it dark matter.

In the Halo

But it is not only in gravitational lenses that dark matter must exist. The way that spiral galaxies rotate and retain their structure over time implies that they are constrained by a gravitational force much greater than their visible matter—their stars and gas clouds—can account for. Each galaxy must be embedded within a vast spherical region, or halo, containing abundant dark matter. The halo, in turn, may be surrounded by an even more extensive sphere, or corona, of dark matter. There could be as much as 10 times more dark matter than visible matter in an average galaxy.

Evidence of dark matter is also provided by galaxy clusters. The galaxies in clusters travel so fast that, in theory, they should rapidly escape from one another. The fact that this doesn't happen suggests that gravity from dark matter must be holding them back.

right
Hot clusters
Part of a cluster containing thousands of galaxies, more than 8 billion light-years away. This is an X-ray image, from a 34-hour exposure by the Rosat astronomy satellite. X-ray images of clusters also reveal pools of very hot gas at temperatures up to 180,000,000°F (100,000,000°C).

Of MACHOs ...

There is circumstantial evidence, then, that visible matter makes up only about 10 percent of the universe, and dark matter 90 percent. But exactly what is dark matter? Because we can't see it or detect it, we don't know for certain what it is. But it could take a variety of forms.

Ordinary visible matter consists of atoms made up of light-weight electrons and a nucleus of heavier particles, protons and neutrons. These heavier particles belong to a class of subatomic particles called baryons; so ordinary matter is called baryonic matter.

The dark matter in the halos of galaxies could be invisible baryonic matter, such as the failed stars we call brown dwarfs, dead stars and black holes. Such invisible objects are called MACHOs (massive compact halo objects).

... and WIMPs

Astronomers also believe that the universe might be flooded with unseen and so-far undetected non-baryonic forms of matter. Physicists have suggested that this matter could consist of particles they call WIMPs (weakly interacting massive particles). WIMPs would be difficult to detect because they would hardly interact at all with ordinary matter. Physicists are currently building detectors deep underground to search for them.

Other elusive particles that we know exist are neutrinos. Neutrinos produced in the Sun's interior are racing through your body right now. They are elusive because they barely interact at all with matter and have no electric charge. Until recently physicists also thought that they had no mass. But research now suggests that neutrinos might have a very small mass—around 1/100,000th the mass of an electron. If this is true, then neutrinos could account for a large proportion of the dark matter in the universe.

Open or Closed?

The amount of dark matter is crucial in deciding the eventual fate of the universe. If there is enough matter, visible and invisible, then its collective gravity will be strong enough to halt the expansion of the universe. If there isn't enough matter, gravity will not be strong enough to halt expansion, and the universe will carry on expanding.

Cosmologists have worked out a critical density for the universe at which gravity will be just strong enough to stop it from expanding. It works out to be a few hydrogen atoms per cubic yard (cubic meter). The ratio of the actual density of the universe to the critical density is designated Ω (omega).

There seem to be three main possible fates for the universe. If Ω equals 1, then the universe will eventually stop expanding, but only after an infinite amount of time, and it will exist forever. This concept is called the flat universe.

If Ω is more than 1, then the universe will eventually slow down and stop expanding. Gravity will then rein in the galaxies and the universe will contract. Over time, everything in the universe will come together in a Big Crunch, a reversal of the Big Bang. This concept is called the closed universe.

If Ω is less than 1, the expansion of the universe that we witness today will continue forever. All the stars, all the galaxies will fade and die; all the black holes will evaporate away. Unimaginably cold, unimaginably vast, the universe will end up as an ocean of subatomic particles with no energy or ability to interact. This concept is called the open universe.

So is the universe flat, closed or open? Will it exist forever, will it end in a Big Crunch, or will it fade away? The jury is still out, but the signs indicate that ours is an open universe. And recent observations suggest that the rate of expansion is actually increasing.

5 | Solar Systems
Other stars have planetary systems too

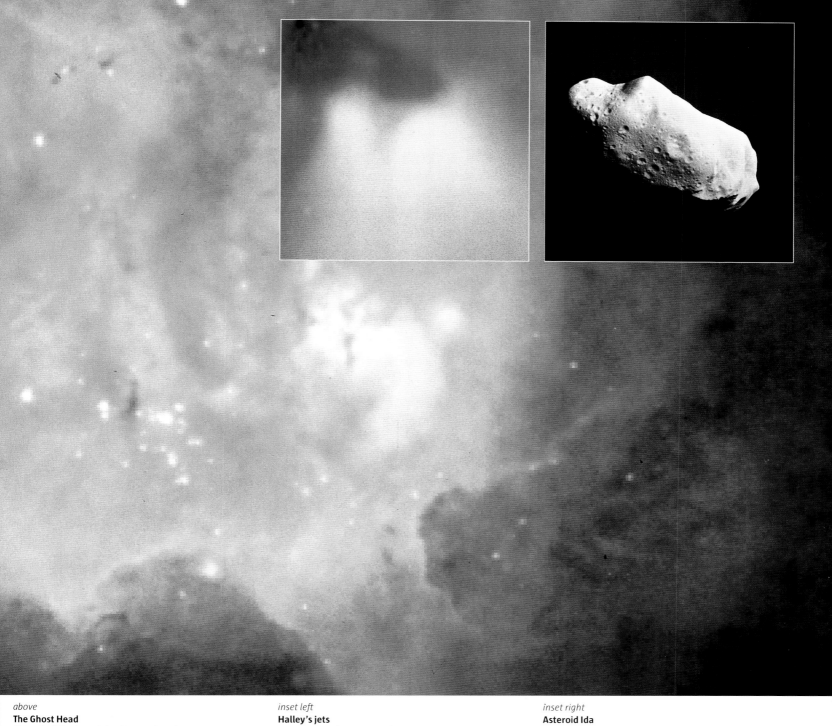

above
The Ghost Head
This colorful celestial cloudscape, found in our galactic neighbor the Large Magellanic Cloud, has been nicknamed the Ghost Head Nebula (NGC 2080). It is one of a chain of star-forming regions around the Tarantula Nebula.

inset left
Halley's jets
Brilliant jets of gas and dust emerge from the potato-shaped nucleus of Halley's Comet. Europe's space probe Giotto returned this image when it had a close encounter with the comet on its last return, in 1986.

inset right
Asteroid Ida
The probe Galileo sent back this image of the asteroid Ida while it was on its way to Jupiter in 1993. Ida measures about 35 miles (55 km) long and, incredibly, has a minuscule moon orbiting around it, named Dactyl.

Planets in the Making

Vast, dark tenuous masses of gas and dust called giant molecular clouds are found throughout interstellar space. They are the stuff that stars are made of, are born from. Five billion years ago, a giant molecular cloud occupied our little corner of the universe. It was from this that the Sun and the planets were born.

Part of the cloud began to collapse under gravity and rotate. As the collapse continued and rotation became faster, the cloud flattened out to form an embryonic solar system that astronomers call the solar nebula. The denser center of this rotating mass began to heat up and glow, becoming the proto-Sun.

Matter continued to feed into the collapsing proto-Sun, which in turn continued to heat up. Eventually, soaring temperatures and pressures triggered nuclear reactions within its core, and it began to shine as a new star. This fledgling Sun then spent roughly 10 million years calming down, until it reached equilibrium between the outward pressure of radiation and the inward force of gravity. It became a stable main-sequence star.

Meanwhile, the surrounding matter had flattened itself into a disk, and it was from this disk that the planets and other members of the solar system eventually emerged.

Scenarios similar to the birth of our solar system are being enacted throughout interstellar space, giving rise to planetary systems around other stars. The first evidence of possible external solar systems came in 1983, when the infrared astronomy satellite IRAS detected dusty disks around the stars Beta Pictoris and Vega. The HST has since spotted dusty disks around a host of stars — dozens in the Orion Nebula alone. They are termed proplyds, for protoplanetary disks, because they are almost certainly planets in the making.

Within our own solar system, how did the planets we know today form out of the primeval dusty disk that girdled the youthful Sun? Surely the planets of other stars would evolve through a similar process.

Inner Planets

In the flattening disk of matter that formed around the proto-Sun, particles began to clump together. Eventually, these clumps grew into masses several thousand yards (meters) across, known as planetesimals. By this time, these great chunks were in orbit around the newborn Sun.

The temperature of the disk of rotating matter varied widely, from a high of around 3,500 degrees Fahrenheit (2,000°C) closest to the Sun to a low of only about –360 degrees Fahrenheit (–220°C) at the outer edge. This variation in temperature dictated the distribution of molecules in the disk.

In the hot, inner region, metals and rock-forming silicate minerals (which have high melting points) collected. Planetesimals with this composition gradually merged by collision to form larger and larger objects, which eventually became the rocky, terrestrial planets — Mercury, Venus, Earth and Mars.

It took roughly 10 million years for these planets to reach more or less their present size, and about another 100 million years for them to sweep up any remaining planetesimals. By this time, powerful blasts of particles in the solar wind had swept the inner solar system clear of gas and volatile substances like water, ammonia and methane.

right
Pictor's planets
Views of the dusty disk around Beta Pictoris from two different HST instruments, the WF/PC (top) and imaging spectrometer. There is evidence to suggest that planets may already be forming inside it.

far right
Pre-collapse cloud
At a distance of only 410 light-years, Barnard 68 is one of the closest dark clouds that will eventually collapse to form stars. Barnard objects contain enough gas and dust to spawn hundreds of stars.

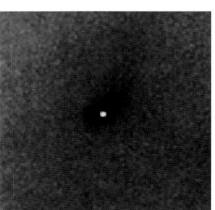

above
Orion proplyds
The HST has imaged
more than a hundred
protoplanetary disks,
or proplyds, in the vast
star-forming region
that is the Orion
Nebula. Clearly, they are
a common by-product
of star formation.

Outer Planets

Planets were also forming in the outer part of the circumsolar disk, where it was cooler. Hundreds of millions of miles away from the Sun, bodies were growing that would become Jupiter and Saturn. They were growing huge by attracting gas, primarily hydrogen and helium, from their surroundings.

Farther out still, the bodies that would become Uranus and Neptune were created when the volatiles, such as water and ammonia, condensed on small, rocky cores. They accumulated relatively little hydrogen and helium, since Jupiter and Saturn had mopped up most of those gases.

Beyond Neptune, ice was the predominant matter. Relatively small ice worlds of ice were formed, of which Pluto was the largest. At the very edge of the solar system matter remained as icy planetesimals just a few miles across. The bulk of these icy bodies remain today. We occasionally see them when they journey in toward the Sun and start to shine—as comets.

Extrasolar Planets

It would be statistically impossible for there not to be other planetary systems like our own. The process of planetary formation in the solar system can't be unique, for this would suppose that the birth of the Sun was a freak occurrence, and we know that this was not the case. And there are all those protoplanetary disks that the HST has spotted that are planets in the making. The question is, how do we prove that there are extrasolar planets—planets beyond our solar system? The answer is: with great difficulty, because planets are tiny compared to stars and don't give off light of their own. Nevertheless, astronomers can still detect them.

The first extrasolar planets were discovered in 1991, not around an ordinary star, but around a pulsar—the remnant of a star that blew itself apart. It was not until 1995 that the first planet was found around an ordinary star, 51 Pegasi. Since then more than 100 extrasolar planets have been discovered. They seem to be roughly the same order of size as Jupiter, but usually orbit much closer to their parent star.

Astronomers have not spotted these planets in telescopes, but have detected them indirectly. The main technique they use is to look for stars with a characteristic "wobble." A wobble, or slight motion to and fro, tends to indicate that one or more planets is orbiting the star and affecting it gravitationally.

Astronomers detect wobbling stars by examining the spectrum of their light. Their motion causes the dark lines in the spectrum of their light to shift, first toward the blue, then toward the red.

If planets are as common as we think, what are the chances of there being planets like Earth? Good. And could such planets harbor life, intelligent or otherwise? Probably. It with this in mind that some astronomers are involved in SETI, the search for extraterrestrial intelligence. The giant radio telescope at Arecibo in Puerto Rico is at the forefront of current SETI research (see page 152).

(see page 152).

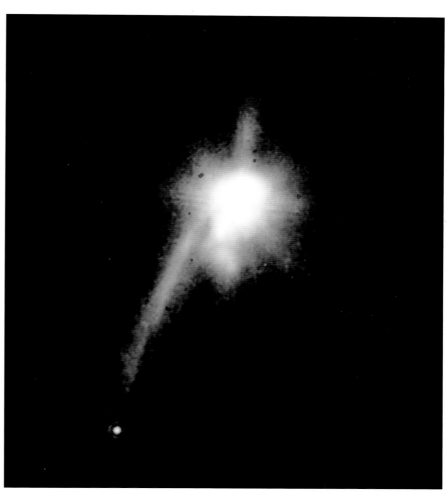

below
A planet pictured?
This infrared image of a close pair of newborn stars reveals a streamer of gas connecting with a faint companion. This small object is too dim to be a star, so could it be a planet ejected from the binary parents?

left
Interstellar frisbees
Looking rather like frisbees, these are some of the protoplanetary disks that have been found in the Orion Nebula. They range in size from about two to 17 times the size of our own solar system.

Family of the Sun

It was the priest-astrologers of the early civilizations in the Middle East who laid the foundations of astronomy. They became familiar with the heavens, but had absolutely no idea of what the universe was really like. They did, however, all agree that the Earth was its center.

The belief that the Earth was flat had, by Aristotle's time (in fourth-century BC), given way to the idea of a round or spherical Earth. Aristotle pointed out that the Earth threw a curved shadow on the Moon. Furthermore, the sphere was the perfect shape and therefore appropriate for a body that was the center of the universe. The dissenting voice of Aristarchus of Samos, a next-generation philosopher who argued that the Earth might orbit the Sun, fell on deaf ears.

Much later, around AD 150, the Alexandrian astronomer Ptolemy elaborated the classical Earth-centered view of the universe, which became known as the Ptolemaic system. According to this system, all the heavenly bodies—Sun, Moon, planets—circled around a central Earth. The stars were fixed to the inside of an all-enveloping dark orb, the celestial sphere, which rotated around the Earth once a day.

This concept held sway throughout classical times and the Dark Ages that followed, when astronomy, like so many other

above
The Copernican system
This old print (1708) depicts Copernicus's Sun-centered universe.

below
The solar system
Orbits of the nine planets in our solar system. The planets are widely scattered, traveling within a region of space 8 billion miles (14 billion km) across. The real edge of the solar system, far beyond Pluto's orbit, is marked by an enormous sphere of cometary bodies called the Oort Cloud.

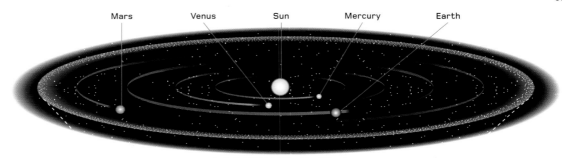

Mars   Venus   Sun   Mercury   Earth

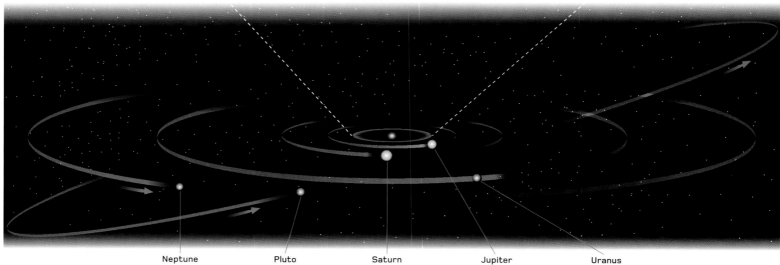

Neptune   Pluto   Saturn   Jupiter   Uranus

branches of learning, suffered almost terminal neglect. Only in the Arab world did the art and craft of astronomy continue to advance. This explains why many stars have Arab names, such as Betelgeuse (in Orion), Algol (in Perseus), and the delicious Zubenelgenubi and Zubeneschamali (Southern Claw and Northern Claw) in Libra. The names of the last two reflect that they were once considered part of Scorpius (the Scorpion).

The Copernican Revolution

In the fourteenth and fifteenth centuries came the rebirth of learning we call the Renaissance. People began to question age-old beliefs, and science (including astronomy) began to advance once more. In 1543, a Polish religious cleric with a passion for astronomy upset the old order and antagonized the Church by putting forward the concept of a Sun-centered, or solar system. His name was Nicolaus Copernicus.

By carefully examining his own and others' observations, Copernicus became convinced that the often bizarre motion of the planets — that they sometimes back-track through the heavens — could better be explained if they and the Earth itself circled the Sun. It took nearly a century for other astronomers, and the Church, to accept the Copernican solar system, which in effect marked the dawn of modern astronomy.

Our Solar System

What is this solar system of ours like? The main bodies are the nine planets and their satellites, or moons. But there are also many other smaller bodies, including asteroids, comets and meteoroids.

The planets circle round the Sun at different distances. They don't exactly "circle" the Sun, but rather travel around it in elliptical (oval-shaped) orbits. With the notable exception of Pluto, all the planets orbit around the Sun in much the same plane (flat sheet) in space. They also all travel in the same direction. If you could view the solar system from a point in space above the North Pole, the planets would orbit the Sun counterclockwise. The planets also have another motion: They spin round in space. Earth spins round once in just 24 hours, but the other planets have different spin times, from under 10 hours (Jupiter) to 243 days (Venus).

There are dwarf planets (such as Earth) and giant planets (such as Jupiter), rocky planets (such as Mercury) and gassy ones (such as Saturn). Diminutive Pluto is in a class by itself; this deep-frozen, icy world, smaller than the Moon, sometimes wanders more than four billion miles (7,000,000,000 km) from the Sun.

What keeps tiny Pluto and all the other planets circling in space? It is the Sun's enormous gravity — enormous because the Sun has such an enormous mass — 750 times more than the mass of all the other bodies in the solar system combined. The Sun thus has much better credentials than the Earth for being the center of our corner of the universe.

top
Rocky dwarf
The Red Planet Mars, one of the inner, terrestrial planets, made up mainly of rock.

above
Gas giant
The beautiful ringed planet Saturn, one of the giant outer planets, made up mainly of gas.

Neighborhood Star

The star at the center of our own solar system, the Sun, is special to us. But in the universe as a whole, it is very ordinary. It is one of as many as 200 billion stars in our Galaxy alone. It may seem big to us, but as stars go, it is a dwarf. The biggest supergiant stars such as Betelgeuse in Orion are hundreds of times bigger in diameter. However, for astronomers, the Sun is of extraordinary importance because it is the only star they can study from close quarters. The Sun is a mere 93 million miles (150,000,000 km) away, while the other stars are light-years from us.

Like other stars, the Sun is made up mainly of hydrogen and helium, the two most common elements in the universe. There are also traces of as many as 70 other elements. All these elements are present as plasma, a form of matter in which the atoms exist as ions (charged particles) in a sea of electrons. Matter takes this form at the kinds of temperatures that exist in the Sun and the other stars.

Solar Energy

The temperature of the bright surface of the Sun, the protosphere, is relatively low, about 10,000 degrees Fahrenheit (5,500°C). As you go deeper into the Sun's interior, temperatures rocket. In the core (center), they reach a high of 30 million degrees Fahrenheit (15,000,000°C).

At such astronomical temperatures and equally astronomical pressures, nuclear reactions take place, which produce the energy that keeps the Sun shining. It has been shining steadily for nearly five billion years, and should continue to shine for at least as long again.

In these nuclear reactions the nuclei (centers) of atoms of hydrogen fuse (join) to form nuclei of helium atoms. In the process small amounts of mass seem to be destroyed, but in fact, they are transformed into energy. Albert Einstein quantified this mass-energy transformation in that most famous of mathematical equations: $E=mc^2$, where E is the energy released when a mass m is transformed, and c is the velocity of light. Since c is a huge value — 186,000 miles (300,000 km) per second — the energy released when even a small amount of mass is converted is incredibly large. In practice, four million tons of hydrogen are converted into energy every second.

right
The Sun's crown
In a total solar eclipse, the Moon moves in front of the glaring surface of the Sun and blots out its light. It is then that we can see the Sun's pearly white outer atmosphere, the corona (crown).

far right
Sunquake
A violent explosion called a solar flare sends shock waves rippling over the surface of the Sun rather like an earthquake does on Earth. So this event is called a sunquake.

far left
SOHO's Sun
Pictured by the SOHO observatory, the Sun is a seething, boiling mass of incandescent gas (mainly hydrogen and helium). The surface looks speckled because of rising (hotter) and falling (cooler) pockets of hot gas.

left
Fiery fountain
Imaged in false color, a tongue of blazing gas erupts from the outer layers of the Sun. Such eruptions, called prominences, reflect the stormy nature of the solar surface.

The Sun's Rays

The energy produced in the Sun's nuclear furnace travels to the surface first as radiation, then on convection currents of rising gas. It pours out from the photosphere as visible light and other invisible radiation, such as gamma rays, X-rays, ultraviolet rays, infrared (heat) waves, microwaves and radio waves. These are different kinds of electromagnetic waves — minute electrical and magnetic disturbances in space. They differ in their wavelengths, with gamma rays being the shortest and radio waves the longest.

The HST can study the light and ultraviolet and near-infrared radiation from the distant stars, but not from the Sun. The intensity of the Sun's radiation would effectively burn out all the sensors. Wisely, HST designers equipped the telescope with a protective sunshade.

Essential Sun	
Diameter at equator:	865,000 miles (1,392,000 km)
Average distance from Earth:	93 million miles (149,000,000 km)
Spins on axis in:	25 days
Mass:	333,000 times Earth's mass
Surface gravity:	28 times Earth's gravity
Surface temperature:	10,000°F (5,500°C)
Core temperature:	30 million°F (15,000,000°C)

left
The far side
The Apollo 13 astronauts snapped this picture of the Moon's far side on their abortive Moon-landing mission in 1970. The dark region is the Sea of Moscow, the only significant sea on the far side.

The Silvery Moon

One of the oddities of the heavens is that, from Earth, the Sun and the Moon appear to be almost identical in size. Nothing could be further from the truth. With a diameter of 2,160 miles (3,476 km), the Moon is actually only 1/400th the size of the Sun. But, because the Moon is 400 times nearer to us than the Sun, the two bodies appear to be the same size.

Of course, the Moon is quite a different body from the Sun, being made of solid rock rather than incandescent gas. And whereas the Sun shines in its own right, the Moon shines only because it reflects sunlight, though not particularly well.

The Moon is Earth's only natural satellite, orbiting our planet about every four weeks. During this time, because of the geometry of the Sun, Earth and Moon in space, we see more or less of the Moon lit up by the Sun. This makes the Moon appear to change shape in the sky — from a thin crescent, to full circle, and back again. These changing shapes, or phases, of the Moon mark one of the great rhythms of nature.

The Lunar Surface

With our eyes, we can make out two general features on the Moon — dark and light areas. The dark areas were once thought to be seas and were named maria (Latin for "seas," singular "mare"). But in reality they are dusty plains. The lighter areas are highland regions, thought to be remnants of the Moon's original crust.

Many of the Moon's mare regions are circular, and this shape suggests their origin. Early in its history, the Moon was pounded by asteroids, and the largest ones gouged out huge basins from the surface. Subsequently, volcanic activity filled the basins with lava, resulting in the dark, flat landscapes that typify the maria.

The fact that the maria were formed more recently than the Moon's highlands is evidenced by the number of craters. On the maria, craters are few and far between, whereas in the highlands they crowd one another. The largest of these craters, such as Bailly and Clavius, measure over 150 miles (250 km) across.

right
Moon rock
A fine-grained volcanic rock brought back from the Moon by the Apollo astronauts. Similar to terrestrial rocks called basalt, it is riddled with holes where gas bubbled out of cooling lava.

right
Taurus-Littrow
Harrison Schmitt examines a huge split boulder near the Apollo 17 landing site at Taurus-Littrow, on the edge of the Sea of Serenity. As ever, the lunar landscape is hauntingly beautiful.

Thanks to close-up investigation by probes and the exciting expeditions by the Apollo astronauts, we know exactly what the Moon is made of. The surface is covered with a kind of soil that crumbles easily, known as regolith. All the rocks are volcanic in origin. The main types are basalt, much like Earth basalts, and breccia, made up of chips of preexisting rocks cemented together.

Some of the moons that orbit other planets resemble our Moon superficially. Many are heavily cratered from heavy bombardment from outer space. But they have a quite different composition. Most of these moons are made up of varying mixtures of ice and rock — except for Jupiter's Io, which has a garish, yellow-orange colored surface of sulfur.

The HST only occasionally targets our own Moon, but not often, because of its relative closeness to the Sun. The HST does, however, routinely target many of the other 60-plus moons of the giant planets.

Essential Moon	
Diameter at equator:	2,160 miles (3,476 km)
Average distance from Earth:	239,000 miles (384,000 km)
Spins on axis in:	27 1/3 days
Circles Earth in:	27 1/3 days
Goes through its phases in:	29 1/2 days
Mass:	1/81 Earth's mass
Surface gravity:	1/6 Earth's gravity

Broom Stars

Of all the objects that light up the night sky, none are as spectacular or as intriguing as comets. Today, when thousands of astronomers are routinely scanning the skies, there is little chance of a comet sneaking up on us undetected, as happened in the past.

In past times, comets seemed to appear suddenly out of nowhere, with their glowing heads and long tails fanning out behind them. To superstitious peoples, the appearance of a comet was a bad omen, a portent of evil, a harbinger of drought or disease, death and destruction. Over 2,500 years ago the Chinese seemed to take a particular interest in comets, which they called broom stars. Ancient Chinese scribes wrote that comets were vile. King Harold of England would probably have agreed, for he was shot in the eye when a comet appeared at the Battle of Hastings in 1066.

A Comet Called Halley

The comet of 1066 was actually making one of its regular appearances in Earth's skies. English Astronomer Royal Edmond Halley was first to recognize this some 600 years later, and so the comet was named after him. Halley's Comet was last seen in Earth's skies in 1986 and will return in about 2061. Thousands of the world's astronomers and a flotilla of spacecraft studied the 1986 appearance of the comet in detail, though the HST had yet to be launched at that time.

Visually, however, the 1986 visit of Halley's Comet was a disappointment. It was faint and difficult to spot among the stars, and observers needed a telescope to see its tail. Comet-watchers had to wait a decade before seeing the first of a pair of spectacular comets with the naked eye. The first was Hyakutake in 1996, the second Hale-Bopp a year later. Hale-Bopp in particular was outstanding—it was one of the great comets of the twentieth century. It hung in northern skies for months, outshining all but the brightest stars. In spring 2002, the fainter Comet Ikeya-Zhang was also easy to see with the naked eye.

below
Hyakutake
Telescope and HST (inset) images of the 1996 comet Hyakutake. Easily visible to the naked eye, it grew a particularly long tail. Hubble images of the comet focused on the region around the nucleus and showed bits that had broken off.

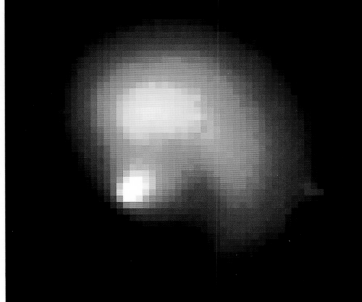

Dirty Snowballs

Astronomers study comets closely, not just because they are so spectacular, but also because they are perhaps the most primitive bodies in the solar system. They seem to originate in a vast reservoir of comets at the outer fringes of the solar system, called the Oort Cloud. This reservoir was created when the planets were being born, and the bodies it contains have remained virtually unchanged since then. They therefore hold the key to understanding the origin and evolution of the solar system.

What are these primitive bodies like? They are made up mainly of a mixture of ice and dust and are often described as dirty snowballs. Far away, in the dark, cold depths of the outer solar system, they remain in deep-freeze. They become visible as comets only when they travel toward the Sun and begin to absorb some of the Sun's warmth.

As they warm up, the icy surface starts to evaporate, releasing clouds of vapor or gas. The gas spurts out in jets, carrying dust with it. The gas and dust form a cloud around the solid part, or nucleus, of the comet. The cloud reflects sunlight, and the comet becomes visible. The cloud can measure up to hundreds of thousands of miles across, but the nucleus itself is usually only a few miles wide.

The pressure of radiation from the Sun forces dust away from the glowing head of the comet, or coma, and shapes it into a fan-shaped, yellowish tail. Magnetic effects associated with the solar wind make the gas particles glow and force them into a second tail called the ion tail (ions are electronically charged particles). In the biggest comets, the tails can stretch for up to 100 million miles (160,000,000 km).

The HST was used to study comets shortly after it was launched and produced spectacular images of Shoemaker-Levy 9 (see overleaf). It also set its sights on Hyakutake and Hale-Bopp. One of its main investigations uses spectrographs to determine the chemical composition of these early relics of our planetary system.

Comet Shoemaker-Levy 9

"I think I've found a squashed comet!" exclaimed US astronomer Carolyn Shoemaker on March 24, 1993. She was examining a photographic plate from a recent exposure during a standard search for comets she regularly conducted at Palomar Observatory in California, with her husband Gene (Eugene) Shoemaker and their colleague David Levy.

In turn, the other two checked the film, they saw a strange apparition with a bar-shaped nucleus, surrounded by a series of little tails and two "wings" of light. None of them had ever seen anything like it, and they were the experts. Together they had already discovered eight comets.

Levy called his friend Jim Scotti at Kitt Peak Observatory in Arizona, who directed his Spacewatch telescope to the position of Carolyn's "squashed comet." Scotti's telescope revealed that it was actually a string of individual comets traveling close together, each with little tails. They were almost certainly pieces of a larger body that had broken up.

The comet-watch trio reported their find, which became known officially as Periodic Comet Shoemaker-Levy 9 (SL9).

In the weeks that followed, astronomers around the world trained their telescopes on SL9. Tracing its path through the heavens revealed that the comet was in orbit around Jupiter. Calculations showed that it must have been the giant planet's powerful gravity that had ripped the larger body apart around July 8, 1993, when it had passed within 13,000 miles (21,000 km) of Jupiter's cloud tops.

Soon, images from a bevy of telescopes, including the HST, showed 21 cometary fragments stretched out in formation — a celestial string of pearls. They seemed to be up to about two miles (3 km) across.

On July 16, 1993, SL9 reached apojove, the most distant point in its orbit around Jupiter, and began to home in on the planet. But calculations showed that this time SL9 would not skim over Jupiter's cloud tops. It would surrender completely to Jovian

above
String of pearls
In May 1994, two months before impact, the 21 icy fragments of Shoemaker-Levy 9 (SL9) are strung out across more than 700,000 miles (1,100,000 km). The HST recorded this image in red light.

above inset
Fireballs
The successive impacts of the comet fragments created huge fireballs in Jupiter's atmosphere, which telescopes on Earth could picture in infrared light.

gravity and impact the planet. Astronomers estimated that the first fragment would hit the surface of Jupiter in July 1994.

Sure enough, on July 16, 1994, fragment A plunged into the Jovian atmosphere, creating a fireball that rose 600 miles (1,000 km) above the cloud tops. One by one, over the next seven days, the fragments bombarded the giant planet, creating fireballs and staining the atmosphere with dark clouds.

Most spectacular was the impact of fragment G on July 18, 1994, which sent a fireball leaping 2,000 miles (3,000 km) beyond the cloud tops. It is estimated to have released the energy equivalent to six million megatons of TNT, which made it three billion times more powerful than the atomic bomb that destroyed the Japanese city of Hiroshima in World War II.

This was the first time in history that astronomers had witnessed such a cosmic collision, and they were ecstatic. But it brought home to humanity just how vulnerable planets, including Earth, must be to bombardment from outer space.

left
G impact
This mosaic of images shows the development of the impact site of SL9's G fragment. Just visible in the first image (lower left) is a tiny plume, marking the moment of impact. The second image shows the impact site 90 minutes later. The third shows the site three days later, along with the site of the L fragment impact. The fourth shows the same region two days later still.

Shooting Stars and Miniplanets

The Earth is under constant bombardment from outer space, as a night of stargazing should demonstrate. From time to time you may see bright streaks in various parts of the sky. It looks as if stars are shooting across the sky or falling to Earth.

However, the shooting or falling stars you see are meteors — bright streaks made by specks of matter burning up in the atmosphere. Interplanetary space is full of specks and larger lumps of rock and metal, particles we call meteoroids.

Some meteoroids have been chipped off of asteroids or even other planets. Others come from comets, which shed material into space on their passage around the Sun. The intense displays of meteors at certain times of the year, known as meteor showers, are associated with the orbits of particular comets.

The meteoroid particles that cause meteors — usually barely larger than sand grains — burn up to ash, which slowly falls and settles on the ground. Occasionally, bigger meteoroid lumps plummet through the atmosphere, to the accompaniment of more spectacular *son et lumière* (sound and light). The "lumière" can be a multicolored fireball, the "son" a muffled sonic boom. Some of the meteoroid survives and falls to Earth, and becomes what we call a meteorite.

Errant Asteroids

If they are big enough, meteorites can blast out mile-wide craters and pose a real threat to life and limb. But asteroids present a much greater threat to humanity.

Asteroids are a group of much bigger lumps of rock and metal, found mainly in a broad band, or belt, between the orbits of Mars and Jupiter. They are also known as the minor planets. But even the biggest, Ceres, is less than 600 miles (1,000 km) wide, and only 25 of the 10,000 or so we know of are bigger than 125 miles (200 km).

The asteroids appear to be the remains of a failed planet, planetesimals that could not unite because of gravitational disturbances, probably principally from nearby Jupiter.

However, not all asteroids orbit within the asteroid belt. Many wander way beyond it, toward Saturn. Others stray in the other direction and wander among the inner planets. Some pursue orbits that take them uncomfortably close to Earth. These are classed as NEOs, or near-Earth objects.

Historically, Earth has almost certainly been hit by asteroids several times. The impact of an asteroid about 10 miles (16 km) wide is thought to have caused a mass extinction of species, including the dinosaurs, 65 million years ago.

below
Blue streak
A curved streak appears on this image of the Milky Way in Centaurus, which looks toward the center of the Galaxy some 25,000 light-years away. It is a trail made by a tiny asteroid just a few light-minutes away.

The chances are that some time, sooner or later, an NEO will target Earth. The hope is that we shall have sufficient warning to try to do something about it. Several organizations around the world now actively look for NEOs. One is the University of Arizona's Spacewatch, with facilities at Kitt Peak Observatory. Discoveries made by such organizations regularly spark off media hype that "the end of the world is nigh."

Close calls

Convincing evidence that Armageddon by asteroids is not just science fiction came within the span of a few weeks in the summer of 2002. First, a NEO (2002 MN), 300 feet (91 m) in diameter, was spotted close to Earth, but receding into space. That was the good news. The bad news was that, three days before, it had missed Earth by less than 80,000 miles (120,000 km).

Not long after, another asteroid (2002 NT7) was spotted much farther away, but it appeared to be making a beeline for Earth, to impact on February 1, 2019. However, subsequent refinement of its orbit suggested that it posed no immediate threat.

Then, a few weeks later, NEO 2002 NY40 flashed past Earth with a close approach of about 300,000 miles (500,000 km). Seen with powerful binoculars, it provided visible evidence that the threat from space is chillingly real for us on Earth.

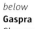

below
Gaspra
Close up, most asteroids are irregular in shape, like this one, Gaspra. It was the first asteroid imaged at close quarters, by the Galileo probe in 1991. About 12 miles (19 km) long, it is one of the rocky asteroids.

below
Varied Vesta
The asteroid Vesta has been studied closely by the HST. Vesta is the third largest asteroid, measuring about 340 miles (550 km) across. The color map (bottom) shows the different composition and varied nature of its surface.

PROFILING:

Eros

Dutch astronomer Carl Gustav Witt discovered asteroid number 433 in 1898. It was called Eros for the son of the Greek love goddess Aphrodite. (In Roman mythology, he was called Cupid and was the son of Venus.)

Eros was the first asteroid found to travel mainly inside the orbit of Mars. It is one of the largest NEOs but fortunately doesn't come much closer to Earth than about 14 million miles (22 million km). Historically, astronomers used observations of Eros early last century to calculate the first really accurate value for the astronomical unit—the distance between Earth and the Sun.

above
The approach
A montage of images taken over three weeks in January and February 2000 as NEAR-Shoemaker homed in on asteroid 433–Eros.

left
Two hemispheres
Two opposite hemispheres of Eros are shown in this pair of images, taken from about 220 miles (350 km). At the left of the top image is the saddle-like feature Himeros. At the center of the lower image is the large crater Psyche.

Even from Earth, astronomers can figure out that Eros is an elongated body that rotates every five hours or so—they can tell this from the way its brightness varies. And they find that it circles round the Sun every 642 days. Eros is one of the most common kinds of asteroids, an S-type; the S standing for silicates, the main mineral ingredients of such asteroids.

We now have a wealth of other data about Eros, for it has been the focus of one of the most incredible feats of interplanetary navigation ever. A spacecraft has not only flown past it, but has gone into orbit around it and then landed. This spacecraft was NEAR (Near Earth Asteroid Rendezvous Mission).

NEAR was launched in February 1996 and took four years to reach its target. Its trajectory took it several times around the Sun and also included a gravity boost from Earth in January 1998. After a failed attempt at orbiting Eros in December 1998, NEAR achieved success on February 14, 2000. It was appropriate that on distant Earth it was the day for lovers—Valentine's Day. Once in orbit around Eros, the spacecraft was renamed NEAR-Shoemaker, for the US planetary geologist and comet-hunter Eugene Shoemaker (see page 124).

NEAR-Shoemaker spent almost exactly a year orbiting Eros at a height of about 30 miles (50 km). It took thousands of close-up images of the asteroid, which proved to be elongated as expected and measured about 21 miles (33 km) long and 8 miles (13 km) across. As well as having a camera for imaging, the spacecraft carried spectrometers to measure the asteroid's composition and a laser range-finder to map its topography.

Like other asteroids that have been imaged, Eros is pockmarked with craters, though it also has smooth regions as well. One of the largest craters, Psyche, is more than 3 miles (5 km) across. Another prominent feature, Himeros, has a saddle shape. Much of Eros is criss-crossed with linear features that were probably created when the tiny world shuddered under impacts with other asteroids. It seems to be covered in a kind of dusty soil, or regolith, presumably produced when the surface rocks are pulverized by meteorite impacts.

During its year-long orbit of Eros, NEAR-Shoemaker greatly exceeded expectations, and its two billion-mile (3,200,000,000 km) mission ended in triumph when it landed on the asteroid's surface on February 12, 2001—something it hadn't been designed to do. After four-and-a-half hours of de-orbiting and braking manoeuvers, it touched down gently at a sedate four mph (6 km/h).

As the spacecraft was descending, it snapped close-up pictures of the surface from altitudes down to 400 feet (120 m), imaging details as small as half-an-inch (1 cm) across. Mission controllers maintained communications with NEAR-Shoemaker for more than two weeks afterward. Said mission director Robert Farquhar in a masterly understatement: "Things couldn't have worked out better."

left
Bizarre grades
The elongated shape, density and spin of Eros combine to create curious gravitational topography—uphill and downhill. In this computer model of the asteroid, for example, red areas are uphill and blue areas, downhill.

left
Inside Psyche
The floor of the crater Psyche, pictured from 30 miles (50 km) away. The surface has a thin coating of soil, or regolith, and is peppered with tiny craters.

left
Landing mode
NEAR-Shoemaker snapped this image from a range of only 800 feet (250 meters) as it was descending to the surface. The large rocks shown are just a few feet high.

6 | The Heavenly Wanderers
The HST keeps a watchful eye on the planets

above

Saturn spectacular

Five pairs of images capture the movement of some of Saturn's moons in the vicinity of the ring system. Ring features visible here, from the planet outward, are the C ring, the Cassini Division and the F ring.

inset left

Stormy Mars

Clouds of dust fill canyons near the equator in this image of Mars taken prior to the Pathfinder landing in summer 1997. Clouds of water-ice cover much of the north polar region.

inset right

Volcanic Io

Jupiter's moon Io is one of the strangest in the solar system because of its abundance of active volcanoes. They spew out sulfur to create a colorful landscape that has been likened to the surface of a pizza.

Scorched Worlds

On many evenings of the year, we can see a bright star hanging in the darkening western sky. We call it the evening star. But star it is not—it is the planet Venus.

Of the five planets we can see in the night sky with the naked eye, Venus is the easiest to recognize. Of the others, Jupiter and Mars are the most distinctive, while Mercury is difficult to see because it always remains close to the Sun, and Saturn tends to merge into the stellar background most of the time. The three other planets—Uranus, Neptune and Pluto—are too faint to be visible at all.

Our knowledge of the planets has been painstakingly gathered over centuries by naked-eye and telescopic observations and more recently by spacecraft, particularly those that have probed deep space to explore planets and their moons, comets and asteroids at close quarters.

Whereas visits by space probes are fleeting, the HST can keep a constant eye on our planetary neighbors and is always on hand to record transient and unexpected phenomena.

Ironically, observing nearby planets with the HST is a lot trickier than observing the far-distant stars. This is because the planets are moving targets, forever changing their position against the background of "fixed stars" as they orbit the Sun.

As you might expect, the two rocky planets that orbit the Sun at a closer range than Earth—Mercury and Venus—are much warmer. Temperatures on Mercury peak at over 800 degrees Fahrenheit (430°C), and those on Venus can rise as high as 900 degrees Fahrenheit (480°C).

Mercury is so hot because it has a very slow rotation period, taking 59 days to spin around once, compared with Earth's rotation period of just 24 hours. This slow spin, coupled with the planet's 88-day movement around the Sun, means that a point on Mercury's surface stays exposed to 176 days of sunshine at a time. This is what makes temperatures soar. On the other hand, when the Sun sets at a point on Mercury, a night begins that lasts for 176 days. Then, temperatures plummet as low as −290 degrees Fahrenheit (−180°C).

Mercury is too small to have enough gravity to hold onto an atmosphere. This lack of atmosphere not only allows the temperature extremes between day and night, but also makes the planet vulnerable to meteorites. In the distant past, lumps of rocky debris rained down on the planet and gouged out craters large and small. Today the craters remain more or less unchanged, for little erosion takes place on this airless world.

Ground-based telescopes show few details of Mercury's surface. Most of our knowledge about the planet has come from Mariner 10, a probe that made three passes of the planet beginning in March 1974. The HST does not attempt to target Mercury because it is always too close to the Sun.

right
Pockmarked Mercury
This mosaic of images of Mercury returned by Mariner 10 in 1974 shows a planet almost completely covered with craters. Most are billions of years old, dating back to a time of intense meteorite bombardment in the solar system.

left
Lunar vistas
Close up, Mercury's surface resembles some of the more heavily cratered regions of the Moon. A major difference between the two bodies is that Mercury doesn't have any large plains like the lunar seas.

Essential Mercury	
Diameter at equator:	3,032 miles (4,880 km)
Average distance from Sun:	36 million miles (58,000,000 km)
Spins on axis in:	58 days 15.5 hours
Circles Sun in:	88 days
Moons:	0

right
Caloris Basin
Circles of mountains ring the biggest feature on Mercury, the Caloris Basin. This was created billions of years ago when an asteroid-sized rock, maybe 60 miles (100 km) across, smashed into the planet.

The Planet from Hell

The planet we know as the evening star, Venus, is rocky like Mercury, but could hardly be more different in most other respects. It is much larger than Mercury, and only a few hundred miles smaller across than our own planet. And like Earth, Venus is surrounded by an atmosphere.

Venus's atmosphere is much thicker than Earth's, with a crushing pressure that is 90 times what we experience. The Venus atmosphere is made up almost entirely of carbon dioxide, the gas that on Earth traps solar energy and causes global warming. On Venus, there is so much carbon dioxide in the atmosphere that it creates a runaway greenhouse effect that has heated up the whole planet to such a staggering temperature, twice as hot as your average domestic oven. It is a temperature that would melt metals such as tin and lead.

The thick clouds that fill the atmosphere permanently hide the planet's surface from our view. However, they are not made up of water droplets like Earth's clouds, but are made up of droplets of sulfuric acid. What a hell of a planet Venus is. If you were to go there, you would be simultaneously oven-baked, crushed and etched to death!

Shaped by Volcanoes

What does Venus look like beneath the clouds? The Venera 9 space probe sent back the first surface picture when it landed in 1975, showing a landscape strewn with scattered rocks.

right
Lifting the veil
The Magellan probe used radar to penetrate Venus's cloud cover and image the surface. The bright region in this global mosaic of images is the larger of the planet's two continents, Aphrodite Terra.

below left
Soaring volcanoes
Maat Mons is one of the most impressive of Venus's many volcanoes. Around 4 miles (6 km) high, it is named for an ancient Egyptian love goddess. Most features on the planet have been given female names. For example, there is a plateau named Guinevere, a chasm named Diana and a crater named Cleopatra.

Radio telescopes operating in radar mode began to reveal the general topography of the planet at much the same time. Radar can reveal surface features because it uses radio waves, which can penetrate clouds.

It was a radar probe named Magellan that completed the first high-resolution survey of Venus's surface between 1990 and 1994. Orbiting around the planet, it returned amazing images, revealing that virtually the whole planet has been shaped by volcanic activity. Thousands of volcanoes cover the landscape, many of them miles high and many almost certainly still active.

The low-lying plains that make up more than four-fifths of the surface of Venus have been formed by repeated flows of lava spewed out from the volcanoes. Most of the volcanoes are of the shield type found widely on Earth, which have a particularly runny lava that can flow a long way.

There are only two main highland regions on Venus, which we can liken to continents. The largest, Aphrodite Terra, lies near the equator and is about as big as Africa. The other, Ishtar Terra, is about Australia's size and lies to the north.

Venus, like Mercury, always stays relatively close to the Sun. The HST can make only occasional observations of the planet, for just a few weeks of the year when it is at its greatest elongation (farthest from the Sun).

The HST can't observe the surface of Venus because it doesn't operate at microwave or radar wavelengths. It can, however, reveal features of the atmosphere by observing at ultraviolet wavelengths.

Essential Venus	
Diameter at equator:	**7,521 miles (12,104 km)**
Average distance from Sun:	**67 million miles (108,000,000 km)**
Spins on axis in:	**243 days**
Circles Sun in:	**224.7 days**
Moons:	**0**

The Red Planet

After Venus, two planets vie for the title of the brightest: Mars and Jupiter — the next two planets beyond Earth, going away from the Sun. They can both shine more brilliantly than any star, but we can easily distinguish them. Whereas Jupiter shines a brilliant white, Mars has a distinct reddish-orange hue, which accounts for its popular name, the Red Planet.

It was the color of Mars that prompted its name in classical times. Likening its color to that of blood and fire, the Romans named it after their god of war.

Through a telescope you can see vague, dark features on Mars, and there are what look like ice caps that come and go at the poles. Careful observation reveals that the planet rotates in a little over one day. Space probes have revealed clouds in the Martian atmosphere.

An atmosphere, clouds, a day of similar length to our own, ice at the poles, dark markings (could they be vegetation?). Surely this must be a planet like Earth? Maybe it also has intelligent life? Many people a hundred years ago believed this to be the case.

Prominent among these believers was US astronomer Percival Lowell, who founded the Lowell Observatory in Arizona specifically to study Mars. He was convinced that the "canals" he thought he saw on the surface were being built by a dying race of Martians, trying to channel water from the ice caps to farmland near the warmer equator.

The reality is quite different. A series of space probes from 1965 to the present day — Mariners, Vikings, Pathfinder, Global Surveyor, Mars Odyssey — have shown Mars to be a barren

left
In retreat
The extent of the polar ice caps on Mars changes with the seasons (which are nearly twice as long as Earth's). These HST images show how the northern ice cap shrinks over a period of five months, between early spring and early summer.

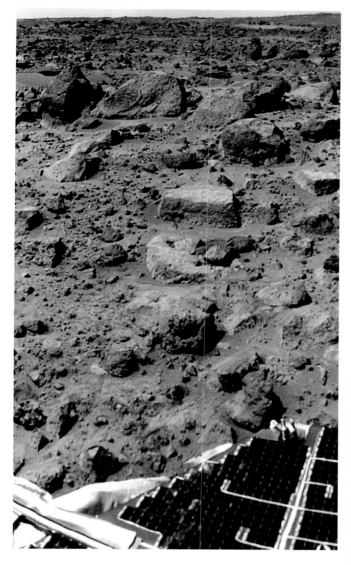

left

Opposition 1999

Varied views of Mars, taken by the HST during the favorable opposition of April/May 1999, when the planet was only about 54 million miles (87 million km) from Earth. Each image depicts the planet as it completes one quarter of its daily rotation.

right

Red rocks

The ruddy appearance of the Red Planet through telescopes is due to the rust-red color of the planet's surface. Here we see the surface close up in an image returned by the Pathfinder lander in 1997. The lander touched down in what is believed to be the estuary of a long-evaporated Martian river.

far right

Getting closer

The opposition of summer 2001 was the closest since 1988, with Mars coming within 43 million miles (68 million km). This image reveals that at the time many dust storms were sweeping round the planet. One is churning over the north polar ice cap (at top); another around Hellas (lower right).

Water and Weather

There may be no canals on Mars, but there are channels that look much like dried-up riverbeds on Earth. They were almost certainly created by flowing water long, long ago. We know that today the only water on Mars is in the form of ice, in the polar caps and the crystals that make up the clouds. Recent spacecraft like Mars Odyssey have also detected the presence of vast quantities of water ice just below the surface — a kind of Martian permafrost.

If water did once flow on Mars, and astronomers are convinced that it did, then the planet must once have had a much milder climate than it does now — one in which water could exist as a liquid. If so, could life have gained a toehold in that milder climate? Maybe. Perhaps the first human explorers of the Red Planet will find out by unearthing fossils of ancient Martian life. We shall have to wait and see.

The HST can shed little light on the prospects of life on Mars. But it does provide a wealth of information about the Martian atmosphere and weather, closely monitoring the changes over the seasons of the Martian year, which is nearly twice as long as our own.

planet, covered in vast deserts and pockmarked with craters. There is only a trace of atmosphere, and temperatures are generally much lower than those on Earth; they rise above freezing only near the equator in mid-summer. There are no signs of canals, or vegetation or life of any kind — intelligent or otherwise.

Mars's Grand Canyon

Mars does, however, boast a natural linear feature — a great gash in the planet's crust that runs along the equator. This ancient geological fault system in the planet's crust is named Valles Marineris, or Mariner Valley, after the Mariner spacecraft that discovered it. It begins near the other outstanding natural feature on Mars — the bulge of the Tharsis Ridge, which carries four of the biggest volcanoes we know in the solar system.

Valles Marineris runs for nearly 3,000 miles (5,000 km), in places widening to more than 250 miles (400 km) and plunging as deep as 4 miles (6 km). Well is it nicknamed Mars's Grand Canyon, but it dwarfs Arizona's famous feature.

Essential Mars	
Diameter at equator:	4,220 miles (6,792 km)
Average distance from Sun:	142 million miles (228,000,000 km)
Spins on axis in:	24 hours, 37 minutes
Circles Sun in:	687 days
Moons:	2 (Phobos and Deimos)

Gigantic Jupiter

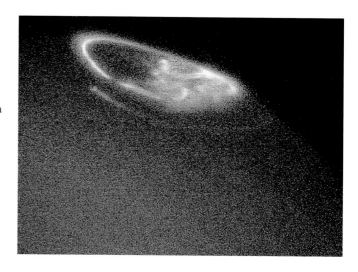

left
Changing spots
The HST keeps a close eye on giant Jupiter and images the hectic activity taking place in the colorful cloud bands superbly. It has in particular been monitoring the marked changes in appearance of the planet's super-hurricane, the Great Red Spot.

right
Jupiter's lights
When charged particles get shaken out of Jupiter's magnetic field and spiral into the atmosphere, they cause displays of aurora, primarily in polar regions. This is similar to what happens in polar regions on Earth, where we see the shimmering light displays of the aurora borealis (in the north) and the aurora australis (in the south).

With 11 times the diameter of Earth and with more mass than all the other planets put together, Jupiter is truly gigantic. Even though it never gets closer to us than about 400 million miles (600,000,000 km), it shines brilliantly in the night sky. This is due not only to its enormous size, but also to its cloudy atmosphere, which reflects light well.

Jupiter is quite a different body from the terrestrial planets. Its atmosphere is much deeper and is composed mainly of hydrogen, with some helium. There is no solid surface, no rocky landscape, beneath Jupiter's atmosphere.

Pressures at the foot of the atmosphere rise so high that they compress the hydrogen gas into liquid, causing the whole planet to be covered by an endless ocean of liquid hydrogen. More than 12,000 miles (20,000 km) below the ocean surface, pressures soar so high that they compress the very atoms of hydrogen. The hydrogen turns into a kind of liquid metal, rather like the liquid metal mercury we find on Earth.

Saturn has a similar makeup to Jupiter, and both planets are often referred to as gas giants. So are the next two planets out, Uranus and Neptune, though they have a somewhat different composition. All four planets have only a relatively small core of rock at the center.

The Stormy Atmosphere

Jupiter has the most fascinating atmosphere of all the planets. It is criss-crossed with bands of colorful clouds, and many other kinds of features—swirls, eddies, plumes and ovals. Astronomers call the dark bands belts and the light ones zones.

The clouds have been drawn out into these parallel bands by Jupiter's rapid rotation. Strangely, the biggest planet rotates the fastest, spinning round once in space in a little under 10 hours. The various bands don't all travel at the same speed; those in the equatorial region travel fastest. And they don't even travel in the same direction—some travel east, some west. The resulting interaction between them sets up violent eddies and turbulence that creates the wavelike features and ovals that cover Jupiter's disk.

Mostly, these features are constantly changing and transient. But one persists: the famous Great Red Spot (GRS). First glimpsed more than three centuries ago, the GRS appears to be an enormous and permanent super-hurricane.

The HST studies Jupiter's atmosphere routinely; it is capable of returning high-resolution images every hour and a half. It studies the planet both in the visible spectrum and at invisible wavelengths. Ultraviolet studies reveal, for example, Jupiter's aurora—the light displays in the polar regions, similar to the Northern and Southern Lights we see in polar regions on Earth.

The Galilean moons of Jupiter—the big four of at least 16 Jovian satellites—also come under the HST's close scrutiny. A prime target is Io, often called the pizza moon because of its colorful appearance. Io is the most dynamic moon in the solar system. Its surface is undergoing constant change due to erupting volcanoes. These volcanoes do not spew out molten rock like volcanoes on Earth, but molten sulfur instead.

left
Io in eruption
Sulfur fumes and dust shoot out of one of Io's many volcanoes, rising to an altitude of 100 miles (160 km) or more. Io's crust is constantly in turmoil because of the constant gravitational tugs this way and that of Jupiter and the large neighboring moons.

below
The big four
The HST has taken a "family portrait" of Jupiter's four large Galilean moons. Ganymede is the largest—indeed, it is the largest moon in the solar system, larger even than the planet Mercury. Callisto is only slightly smaller.

Ganymede

Callisto

Io

Europa

Essential Jupiter	
Diameter at equator:	88,850 miles (143,000 km)
Average distance from Sun:	483 million miles (778,000,000 km)
Spins on axis in:	9 hours, 55 minutes
Circles Sun in:	11.9 years
Moons:	16+

The Ringed Wonder

Encircled by a set of bright shining rings, the gas giant Saturn remains a firm favorite among astronomers. The other gas giants (Jupiter, Uranus and Neptune) also have ring systems, but they are but a pale imitation of Saturn's and cannot readily be seen from Earth.

With a similar makeup to that of Jupiter, Saturn has a banded atmosphere, but the parallel cloud bands are much less distinctive. Not so much activity takes place in the atmosphere, though furious storms do rage from time to time.

Ring-a-Ring

Saturn's rings reflect light brilliantly and make the whole planet much brighter than it otherwise would be. In the night sky, it can outshine all the stars except Sirius and Canopus, but for only some of the time. Even when it is farthest away from us — at a distance of nearly 900 million miles (1,400,000,000 km) — it is still as bright as a first-magnitude star.

The amount of extra brilliance provided by the rings varies. Because of the inclination of Saturn's spin axis, on Earth we see the rings at different angles during the almost 30 years it takes Saturn to orbit the Sun. Roughly every 15 years, the rings all but disappear, when they appear edge-on to our line of sight. The last ring-plane crossing occurred in 1995.

Rings and Shepherds

In telescopes, we can make out three main rings from Earth: the A, B and C rings, from outer to inner. Brightest is the B ring, which is separated from the A ring by the apparently ring-less

below
Changing aspects
Saturn's incomparable rings present different aspects to us year by year. These HST images were taken between 1996 (bottom), a year after the ring-plane crossing, and 2000 (top), when the rings were nearly fully open.

Cassini Division. In total, the ring system measures some 170,000 miles (272,000 km) across.

Spacecraft, particularly the two Voyagers, have changed our perspective on the rings. They have spotted more rings inside and outside those visible from Earth. It turns out that each set of rings is made up of thousands of individual ringlets. The ringlets are our perception of particles of material whizzing in orbit around the planet at high speed. These particles are made up mainly of ice, which is why they reflect light so well. Some particles are as small as sand grains; others are as big as boulders.

The camera eyes of the Voyager probes also provided an explanation of why the particles remain in the rings. They detected a number of tiny moons among the rings that by their gravity seem to keep the ring particles in place. They are called shepherd moons, because they seem to "herd" the particles like a shepherd herds sheep.

As with Jupiter, HST observers concentrate on monitoring the changes in Saturn's atmosphere. They have logged the progress of many a storm, after being alerted by astronomers on the ground. This is an example of the growing cooperation between terrestrial and space astronomers.

The HST also sets its sights on Saturn's planet-sized moon Titan. Bigger than Mercury, Titan is unique among moons because of its thick atmosphere, which astronomers believe might resemble Earth's early atmosphere in its composition.

above
Voyager's Saturn
One of the stunning images of Saturn Voyager 2 returned when it flew within 78,000 miles (124,000 km) of the ringed planet in 1981. The rings are amazing to behold.

left
Titan's surface
The HST has peered through the thick, cloudy atmosphere surrounding Saturn's largest moon, Titan, to glimpse details of its surface. It is thought that the surface might be covered with methane ice and snow.

right
Titan's atmosphere
Haze in Titan's atmosphere (blue), which is made up mainly of nitrogen, like Earth's. Surprisingly, the pressure of the atmosphere is 50 percent higher than it is on Earth.

Essential Saturn	
Diameter at equator:	74,900 miles (120,500 km)
Average distance from Sun:	888 million miles (1,429,000,000 km)
Spins on axis in:	10 hours, 39 minutes
Circles Sun in:	29.5 years
Moons:	23+

right
False-color Saturn
It took the HST 20 exposures with the WF/PC over a period of 20 hours to produce this full image of Saturn. The rings are lit by the Sun from underneath, and it is the least dense sections of the rings that let most light through. The brightest features we see, therefore, are the outer F ring, the Cassini Division and the inner C ring.

left
Infrared Saturn
To celebrate the HST's eighth birthday, the science team "gift wrapped" everybody's favorite planet in vivid colors. It was imaged in the infrared by NICMOS—the first time this instrument had been turned on the planet.

right
Ring-plane crossing
Two images of Saturn taken three months apart. The upper image (August 1995) records the ring-plane crossing—the time when the plane of the ring system crosses our line of sight; the next crossing will take place in 2010.

left
Saturn's lights
Saturn's polar auroras show up brilliantly in ultraviolet light. They are just as spectacular as Jupiter's.

New Worlds

On the night of March 13, 1781, William Herschel was studying the stars in Gemini at his home in Bath, England. German-born, he had neglected his profession as a musician to become an enthusiastic astronomer. Among Gemini's stars he came across an object that intrigued him. "A curious nebulous star or perhaps a comet," he recorded. But this body was no star or comet; it was a new planet, Uranus.

When astronomers determined the planet's orbit, they were astonished. This new wandering star pursued an orbit that took it 1,800 million miles (2,900,000,000 km) from the Sun. This was twice as far away as Saturn, regarded since ancient times as the farthest planet. Herschel's discovery literally doubled the size of the known solar system, and triggered the search for more new worlds.

Small Giant

Uranus is the third-biggest planet after Jupiter and Saturn, but it is only about a quarter as wide as Saturn. The strangest thing about the planet is that its spin axis is tipped right over. The axis of spin for all the other planets is more or less at a right angle to the plane of its orbit around the Sun. But Uranus's spin axis is nearly in the orbital plane itself.

In appearance, Uranus is a featureless, bluish-green orb. Like the other gas giants, its deep atmosphere is made up mainly of hydrogen and helium. There are also significant traces of methane, the gas we use to cook on Earth. It is this gas that is responsible for the color of the atmosphere, because it absorbs red wavelengths from sunlight, producing a blue color.

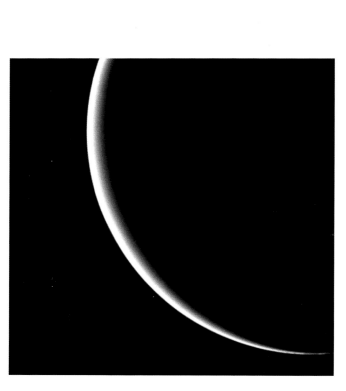

Surprise, Surprise

In 1977, Uranus sprang another surprise. Astronomers setting out to chart the occultation of a star by the planet discovered sets of rings around it. In 1986, Voyager 2 pictured the rings more clearly, showing that there are 11 in all. Narrow and as black as coal, they contrast markedly with Saturn's broad, bright rings.

Another intriguing Voyager discovery came when the probe closed in on Miranda, the smallest Uranian moon we can see from Earth. Miranda's surface displays geological features that are unique in the solar system. Quite different landscapes abut one another with no gradual transition between them.

Essential Uranus	
Diameter at equator:	31,760 miles (51,120 km)
Average distance from Sun:	1,786 million miles (2,875,000,000 km)
Spins on axis in:	17 hours, 14 minutes
Circles Sun in:	84 years
Moons:	17 +

above
Hazy atmosphere
Imaging in the near-infrared, the HST picks out high-altitude haze (pink, dark blue) and lower-level methane clouds (yellow, pale blue) in Uranus's atmosphere.

left
Crescent Uranus
A thin crescent of Uranus, pictured by Voyager 2 as it completed its encounter with the topsy-turvy planet.

Patchwork Landscape

It is possible that such geological patchwork might have come about as a result of a catastrophic collision with another body eons ago. The collision would have smashed the moon to pieces, and then gravity would have caused the moon to reform again, with all the pieces of different geology jumbled together haphazardly.

Voyager 2 also spied a number of tiny moons, invisible to us from Earth. Among them were two shepherd moons that orbit on either side of the outer, Epsilon ring. Named Cordelia and Ophelia, they are both only about 20 miles (30 km) across.

The HST carries out regular observations of Uranus, its rings and its moons. It has spied more activity in the atmosphere than did Voyager 2, following the changing patterns of clouds and haze. Its observations of the planet's moons, particularly the new ones discovered by Voyager, have enabled astronomers to refine their orbits.

above
Epsilon's ringlets
Many individual ringlets make up Uranus's outermost ring, the Epsilon, pictured here by Voyager 2. The Epsilon is also the broadest ring, measuring about 60 miles (100 km) across.

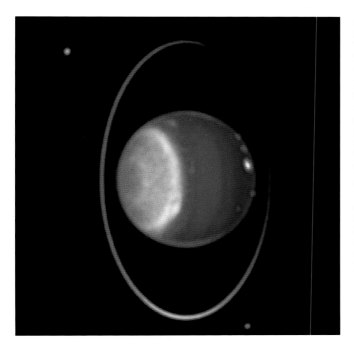

above
Amazing Miranda
The strange patchwork landscape of Miranda, where different kinds of terrain butt together — cratered hilly regions, chevron or V-shaped areas, and oval grooved regions that look like a gigantic racetrack. There is no geology like it anywhere else in the solar system.

left
Clouds and storm
Cloud bands and storms are clearly visible in this infrared HST view of Uranus.

Blue Planet

The enthusiastic search for other planets triggered by Herschel's discovery of Uranus came to fruition in 1846. On September 23 of that year, German astronomer Johann Galle discovered planet number eight, which was subsequently called Neptune. It proved to be only a fraction smaller than Uranus. Recent observations have shown that the two planets are remarkably similar, even though their orbits are more than a billion miles (1,600,000,000 km) apart.

Neptune has relatively more methane in its hydrogen/helium atmosphere, and as a result the planet appears a deeper blue than Uranus. But both planets seem to have a similar ocean of warm water, liquid ammonia and methane underneath the atmosphere.

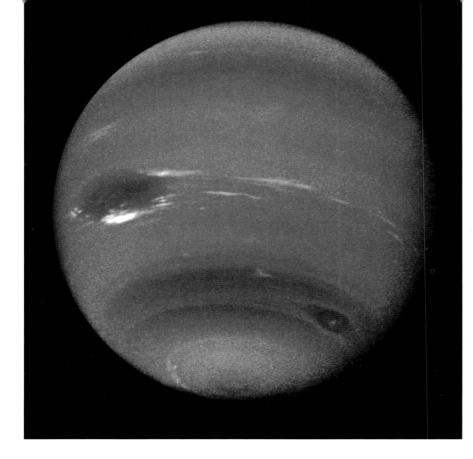

above
The Great Dark Spot
Voyager 2 gave astronomers their first ever close-up view of Neptune, revealing a planet with a remarkably dynamic atmosphere. The Great Dark Spot was the dominant feature. Strangely, it was on a similar scale, and at a similar latitude, to Jupiter's Great Red Spot.

right
Changing weather
The HST monitors the rapidly changing weather on Neptune, tracking the development of the cloud systems in the deep atmosphere. It hasn't yet spotted any storm feature like the Great Dark Spot Voyager 2 found in 1989.

Neptune's Clouds and Storms

When Voyager 2 flew past Neptune in 1989, it revealed that the blue planet's atmosphere was much more active — had more weather — than Uranus's. Being much farther from the Sun, Neptune receives less solar energy, which is usually what drives planetary weather systems. So you might expect that Neptune's atmosphere would have little weather.

The truth is quite the reverse. Voyager's cameras spotted distinct bands in the atmosphere, rather like Jupiter's and Saturn's. There were also oval storm-type features; a particularly large one became known for obvious reasons as the Great Dark Spot. There were also high-level, wispy clouds, rather like the cirrus clouds we get in Earth's atmosphere. But Neptune's wispy clouds seem to made up of crystals of methane rather than water ice.

Why is Neptune's atmosphere more active than that of Uranus? It seems that the planet has an internal heat source, which drives the planet's weather systems. Internal heating would also explain why at the cloud tops, both Uranus and Neptune have more or less the same temperature, of about −350 degrees Fahrenheit (−210°C).

Essential Neptune	
Diameter at equator:	30,780 miles (49,530 km)
Average distance from Sun:	2,800 million miles (4,505,000,000 km)
Spins on axis in:	16 hours, 6 minutes
Circles Sun in:	165 years
Moons:	8

Neptune's Rings and Moons

Like Uranus, Neptune has a ring system, but it is even sparser and fainter. Two of the rings are relatively bright, and the other three very faint indeed. The two brightest are named Adams and Leverrier, after the two mathematicians who independently figured out where the eighth planet could be found (Englishman John Couch Adams and Frenchman Jean Urbain Leverrier).

On its flyby of Neptune, Voyager 2 discovered six new moons, adding to the two that can be seen from Earth, Nereid and Triton. One was Proteus, bigger than Nereid but impossible to see from Earth, because it orbits so close in that it gets lost in the planet's glare. Of the remaining five new moons Voyager discovered, four orbit within the ring system, presumably acting as shepherds for the ring particles.

When the HST first trained its instruments on Neptune, it revealed no trace of the Great Dark Spot Voyager had seen. But it spotted plenty of other cloud activity in the atmosphere.

Ice Worlds

With Neptune, we come to the end of the progression of gas giants that dominate the outer reaches of the Solar System. Then, as we penetrate ever deeper into space, we encounter planetary bodies that we call the ice worlds.

We don't even have to leave Neptune to find one. Neptune's largest moon Triton is one. Measuring some 1,700 miles (2,700 km) in diameter, it is made up of rock and ice. Its surface is covered with frozen gases, such as nitrogen and methane, as well as ice. The most curious feature of Triton's surface is its volcanoes. Totally unlike Earth's molten lava volcanoes, those on Triton emit a kind of slushy, frozen gas-ice in a process termed cryovolcanism. Geyser-like eruptions also puff out gas mixed with dark particles, which form dark streaks on the pale ice and snow.

right
Icy geysers
Neptune's biggest moon Triton is in deep freeze. It has the coldest surface (−390°F/ −235°C) of any body we know in the solar system. The image shows the south polar region, which seems to be covered with pinkish snow. The dark streaks show where fine dust is erupting from icy geysers.

Planet #9

Had planetary circumstances been different, Triton might have pursued an independent orbit round the Sun and become a planet in its own right. This is what seems to have happened to another ice world, smaller than Triton, which we know as the planet Pluto.

Astronomers still continued to try to find new planets long after Neptune had been discovered. Prominent among them was Percival Lowell, noted for his belief in intelligent Martian life. At the Lowell Observatory in Arizona, Clyde Tombaugh eventually found a ninth planet in February 1930. It was named Pluto.

Pluto is the only planet that spacecraft have not visited yet, so we know less about it than the others. It is much too remote for ground-based telescopes to spy surface features. But spectroscopic analysis has shown that it is covered with frozen nitrogen, methane and other gases.

When Pluto is closest to the Sun in its highly elliptical orbit, some of the ice evaporates, giving the planet a faint atmosphere. At such times — for 20 of the 248 years it takes to orbit the Sun — Pluto wanders within the orbit of Neptune. Then Neptune temporarily becomes the farthest planet. Pluto was last inside Neptune's orbit between 1979 and 1999.

Double Planet

In March 1978, Pluto was discovered to be a double planetary system, when US astronomer James Christy discovered another body circling closely around it. Astonishingly, this moon, named Charon, proved to be half the size of Pluto.

Charon is probably similar to Pluto in appearance and makeup, though it does appear to be a little darker and may well have a little more ice and less rock than Pluto.

Triton, Pluto and Charon seem to be variations on the same kind of body, and astronomers know that there are many other

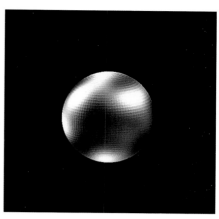

above
Pluto revealed
Pluto is a tough target even for the HST. But the instrument has managed to resolve vague features on the planet's surface, though it is difficult to guess what they might be.

ice worlds like them in the outer solar system. They exist in a region known as the Kuiper Belt, named for US astronomer Gerard Kuiper, discoverer of, among other things, Neptune's moon Nereid. More than 80 Kuiper Belt Objects (KBOs) have been discovered to date. Some appear to be comet-like bodies, and may evolve into comets over time, moving into the inner solar system to reveal themselves in the skies of Earth.

The HST has become a lead instrument in the study of Pluto, able to resolve some surface features. It is also proving a valuable tool for observing KBOs. For example, in summer 2002 its new Advanced Camera for Surveys was able to resolve the disk of a recently discovered KBO, and make the first direct measurement of such a world. Named Quaoar, for a creation god of the Native American Tongva tribe, this body proved to have a diameter of around 800 miles (1,300 km). This made it the biggest body found in the solar system since the discovery of Pluto 72 years before.

above
Close companions
Ground-based telescopes find it hard to separate Pluto and its moon Charon, but the HST does so with ease. Charon is surprisingly big for a moon of such a small planet, and it circles close in, only 12,000 miles (19,000 km) from Pluto.

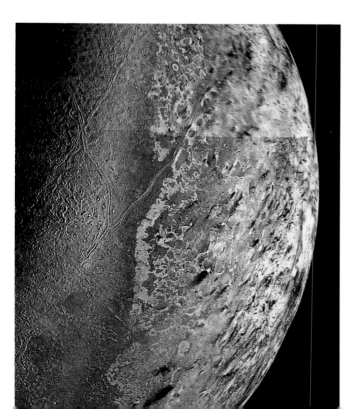

Essential Pluto	
Diameter at equator:	1,429 miles (2,300 km)
Average distance from Sun:	3,676 million miles (5,916,000,000 km)
Spins on axis in:	6 days, 9 hours
Circles Sun in:	247.7 years
Moons:	1 (Charon)

Background Briefing

Introducing Telescopes | The Hubble Space Telescope

above

VLT telescopes

The four 27-foot (8.2-m) reflectors that make up the Very
Large Telescope (VLT) at the Parana Observatory at Atacama
in Chile. The telescopes are named for sky objects in the
native Chilean dialect. In the foreground is Yepun; behind it
(from the left) are Antu, Kueyen and Melipal.

Introducing Telescopes

We can explore the universe with just our eyes and view the constellations, meteors, comets and eclipses. However, as an optical instrument, the eye has a great disadvantage. It has a small aperture (opening)—the pupil—to let the light through. The principal instrument that astronomers use, the telescope, has a much greater capacity for light-gathering.

The Italian genius Galileo was the first to build a telescope and train it on the heavens, in the winter of 1609–10. He became the first to see the mountains on the Moon, the phases of Venus and the four large moons of Jupiter (henceforth called the Galilean moons). Galileo used a set of lenses to gather and focus light from the heavens. Many amateurs still use this kind of telescope, called a refractor. An objective lens at the front gathers and focuses incoming light, and the observer views the image formed through an eyepiece.

However, most astronomers use a telescope with mirrors, called a reflector. A curved, parabolic mirror (the primary) is used to gather and focus the light into an image, which is then viewed through the eyepiece. The instrument favored by amateur astronomers is called a Newtonian reflector, because it uses the same configuration, as the reflector built by Isaac Newton around 1672. A secondary mirror reflects light gathered by the primary mirror into the eyepiece at the side. In an alternative configuration called the Cassegrain, a secondary mirror reflects light into the eyepiece through a hole in the primary. This is the configuration used by the HST.

The progress of astronomy since Galileo's time is a result of the improvement in the size and design of telescopes. The bigger the telescope, the more efficient it is in gathering light, and the more clearly it can resolve (distinguish) objects.

William Herschel built the finest instruments of his day and discovered a new planet, Uranus, in 1781. Lord Rosse built a gigantic instrument he called the Leviathan and was first to spot the spiral nature of M51 (the Whirlpool Galaxy) in 1845. With the groundbreaking 100-inch (2.5-m) Hooker telescope, Edwin Hubble proved in the 1920s that there were star systems—galaxies—beyond our own and that the universe appeared to be expanding.

Very Large Telescopes

Today's foremost instruments use very large mirrors. The two Keck telescopes at the Mauna Kea Observatory in Hawaii have mirrors 33 feet (10 m) across. The four instruments that make up the Very Large Telescope (VLT) in Chile each have mirrors 27 feet (8.2 m) across.

The mirrors in these telescopes are not made of a single piece of glass. It would be difficult to control and adjust a mirror of that size without creating distortions. These mammoth mirrors are made in segments, which merge together to make one big mirror. Each segment is individually supported and controlled by a computer that constantly adjusts it so that, all together, the segments always form the perfect shape. This technique is known as active optics.

Individually, each of the four telescopes of the VLT is superb, and when they work together as they are designed to, they produce spectacular images that match the quality of those from the HST. When they work in unison, along with three smaller 6-foot (1.8-m) instruments, they create an effective telescope 390 feet (120 m) across.

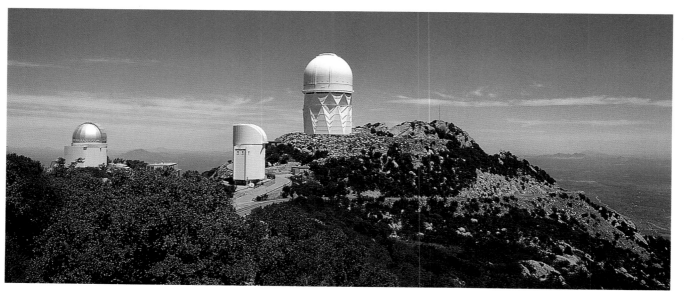

left
Kitt Peak
Telescope domes at Kitt Peak National Observatory (KPNO), near Tucson in Arizona. Kitt Peak is one of the world's finest observatories, with superb seeing conditions thanks to its high altitude and the arid climate.

above
Double cluster
A fascinating double
star cluster (NGC 1850)
in the Large Magellanic
Cloud. It represents a
stellar association not
found in our own
Galaxy. It is interesting
to compare this VLT
image with an HST
image of the same
region on page 71.

left
The Butterfly
This beautiful planetary
nebula of southern
skies is well named the
Butterfly. It was a first-
light image from the
first VLT telescope,
Antu, which became
operational in 1998.

The Invisible Universe

In 1931, a communications engineer named Karl Jansky from Bell Laboratories in New Jersey, was trying to identify interference he was getting on his radio equipment. After ruling out any local sources, he realized what was happening. The interference was coming from the heavens. That observation proved to be the springboard to one of the most exciting branches of astronomy — radio astronomy. It led astronomers to discover whole new kinds of heavenly bodies, including quasars, pulsars and radio galaxies.

Astronomers tune into the radio waves that emanate from the heavens with radio telescopes that look nothing like their optical counterparts. A radio telescope usually takes the form of a huge metal dish. The dish collects incoming radio waves and focuses them on a central antenna. The signals then pass to a receiver. After amplification, they are fed to a computer for analysis and display as a false-color image, with different colors being assigned to different radio intensities and wavelengths.

Big and Versatile
The biggest single radio telescope is located near Arecibo in Puerto Rico. Measuring 1,000 feet (300 m) across, it is built into a natural bowl in a hilltop. Since it is fixed, it must rely on the Earth's rotation to direct it toward different parts of the sky. A much more versatile instrument is the Very Large Array near Socorro, New Mexico. It is made up of 27 separate dishes, each 82 feet (25 m) across. The dishes are mounted on rail tracks and can be moved into various configurations. Together they present a collecting area some 20 miles (30 km) across.

Radio telescopes at different locations can also be linked to simulate even larger dishes. This is called VLBI (very long baseline interferometry). Theoretically, it can create an effective collecting area with the diameter of the Earth.

Invisible Astronomy
The radio waves that pour down on Earth from the heavens come from what is often called the invisible universe — the universe we can't see because our eyes can't detect it. As witnessed by radio images, the invisible universe can look nothing like the visible one. Still, there is more to the invisible universe than just radio waves.

We detect objects in the heavens by the energy they emit. Visible light and invisible radio waves are just two forms of energy we receive from heavenly bodies such as stars and galaxies. They also give off energy as invisible gamma rays, X-rays, microwaves, and ultraviolet and infrared rays. All these rays are kinds of electromagnetic radiation, which differ only in wavelength.

above
Parkes RT
One of the largest radio telescopes in the Southen Hemisphere, at the Australian National Radio Observatory, near Parkes, New South Wales. Its dish measures 210 feet (64 m) in diameter.

In order to get a comprehensive picture of the heavenly bodies, astronomers ideally need to study them at all of the wavelengths they give out. On the ground, astronomers can study light and radio waves from the heavens because they pass through the atmosphere. But they can't study other wavelengths because the atmosphere absorbs them. This is one of the reasons that astronomers launch telescopes and other instruments into space.

Another reason for launching telescopes into space is that, orbiting hundreds of miles high, they are above Earth's polluted atmosphere. The atmosphere, in effect, offers us a dirty window on space. It is full of dust, clouds, water vapor and shimmering air currents, all of which at distort the faint light that reaches us from the heavens.

Space Observatories

Spacecraft have been returning information about the space environment from the very beginning of the Space Age. Launched on January 31, 1958, the first US satellite, Explorer 1, discovered intense donut-shaped bands of radiation around the Earth, which are called the van Allen belts. The first dedicated astronomy satellite, OAO-2 (Orbiting Astronomical Observatory-2) went into orbit 10 years later.

Since that time, scientists have launched dozens of astronomy satellites — space observatories — that span the full range of invisible wavelengths (except for radio waves). From shortest to longest, they cover gamma rays, X-rays, ultraviolet rays, infrared rays and microwaves. The HST, while primarily used for visible light, also covers some ultraviolet and near-infrared wavelengths.

left
Radio galaxy
The powerful radio source 3C-75, a radio galaxy that pumps out millions of times more energy at radio wavelengths than our own Galaxy.

below left
The radio Sun
Like all stars, the Sun gives off radio waves as well as other visible and invisible radiations. This image is a radio picture of the Sun.

below right
IRAS in orbit
High above Earth's atmosphere, astronomy satellites like IRAS can pick up radiations that never make it down to Earth's surface. Infrared satellites like IRAS and ISO have limited life because the supercold liquid helium they use to make their detectors ultrasensitive eventually evaporates.

Gamma Rays

Gamma rays have the shortest wavelengths and pack the most energy. They are given off by the most violent activities that take place in the universe. Scientists have detected them coming from pulsars and quasars and in incredibly energetic bursts from unknown events in distant galaxies. Other gamma-ray sources have been identified as events in which ordinary matter and its mirror-image opposite, antimatter, annihilate each other.

The US Compton Observatory, launched in 1991, provided the first comprehensive view of the gamma-ray universe. Europe's Integral (International Gamma-Ray Astrophysics Laboratory), launched in 2002, is designed to provide unprecedented high-resolution imaging of gamma-ray sources. It operates in a looping orbit that takes it as far as 95,000 miles (153,000 km) from Earth. In this orbit it spends most of the time outside the potentially damaging van Allen radiation belts in a relatively undisturbed space environment that is ideal for long-duration, real-time observations.

X-Rays

The wavelengths of X-rays are longer than those of gamma rays and shorter than those of ultraviolet rays. They still pack a lot of energy, which is why they can penetrate body tissue. In the heavens, X-rays are emitted by very hot objects — with temperatures of millions of degrees. Typical X-ray sources include the solar corona (the outer atmosphere of the Sun), supernova remnants and the gas that spirals around black holes. Key X-ray observatories include Rosat (Roentgen satellite, launched in 1990), Chandra (1999) and XMM-Newton (1999).

Ultraviolet Rays

Ultraviolet wavelengths are just shorter than those of the violet light of the visible spectrum. Some ultraviolet (UV) gets through Earth's atmosphere — it is the radiation that tans us. Thankfully, most UV is absorbed by a layer of ozone in the upper atmosphere. The hottest stars are best observed in the ultraviolet because they give out most energy at these wavelengths. The International Ultraviolet Explorer (launched in 1978) provided outstanding data for 18 years, including the study of Nova Cygni (1978), Halley's Comet (1986) and the naked-eye supernova in the Large Magellanic Cloud (1987).

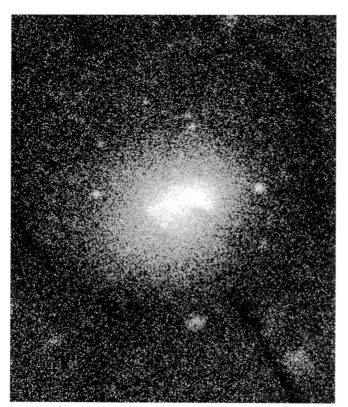

left
Rosat
An X-ray image of the Abell 2256 cluster of galaxies returned by Rosat. It reveals a pool of gas at the center of the cluster with a temperature of up to 180 million°F (100 million°C).

above
Einstein's Andromeda
X-ray sources in the Andromeda Galaxy, imaged by the Einstein Observatory (launched 1978), originally designated HEAO-2. Einstein was the first large X-ray telescope, which pinpointed more than 5,000 sources.

right
UV view
A shuttle-borne ultraviolet telescope called Astro-1 imaged this face-on spiral galaxy, M74, in Pisces. The bright points highlight regions of intense star formation in the spiral arms.

Infrared Rays

Infrared rays have slightly longer wavelengths than visible red light. We receive heat from the Sun is in the form of infrared rays. Some infrared wavelengths reach the ground and can be studied from mountaintop observatories. But most infrared wavelengths are completely absorbed by the atmosphere, principally by carbon dioxide and water vapor. Airborne observatories such as SOFIA (Stratospheric Observatory for Infrared Astronomy) study the infrared wavelengths from high altitudes. But the most spectacular work at infrared wavelengths has been done by satellites such as IRAS (Infrared Astronomy Satellite, launched in 1983) and ISO (Infrared Space Observatory, launched in 1995).

Microwaves

Microwaves have the shortest radio wavelengths. They are used terrestrially to relay communications signals and in cooking with microwave ovens. COBE (Cosmic Background Explorer), launched in 1989, used microwaves to survey the cosmic background radiation of the whole sky. It detected slight "ripples," or variations, in the radiation, regarded as the thermal "echo" of the Big Bang. COBE's findings supported current thinking on the Big Bang and cosmological theory.

Solar Observatories

By studying the Sun, we find out what ordinary stars are like — for there are billions of stars like it in our Galaxy alone. And there are also more pressing reasons for studying this, our local star, because what happens in the Sun can affect Earth and change our "space weather."

As with the other stars, we study the Sun both from the ground and from space. On the ground, astronomers use quite different instruments. Solar telescopes produce images by projection, using mirrors. The McMath solar telescope at Kitt Peak Observatory in Arizona is the world's biggest, with a distinctive triangular design.

The astronauts on the Skylab space station made the first comprehensive study of the Sun from space in 1973–74. Using a series of eight instruments, they probed the Sun at many different wavelengths and achieved spectacular results. More recently, Solar Max (short for Solar Maximum Mission) was launched in 1980 to monitor the Sun at a time of solar maximum, when sunspots and other solar activities reach their climax.

Ulysses (launched in 1990) embarked on a circuitous route via Jupiter to spy on the polar regions of the Sun, which had never before been investigated. SOHO (Solar and Heliospheric Observatory) began continuous observations of our star in 1995. It is "parked" in a gravitationally stable point in space, called Lagrangian Point 2, about 930,000 miles (1,500,000 km) from Earth.

above
ISO's Whirlpool
The famous Whiirlpool Galaxy (M51) in Canes Venatici, imaged by Europe's Infrared Space Observatory (ISO). The observatory returned not only images but also data on the composition of the interstellar medium (see also page 88).

right
Leftovers
A whole sky map produced from data returned by COBE, the Cosmic Background Explorer (launched 1989). The satellite scanned the sky at microwave wavelengths to monitor the faint radiation left over from the Big Bang.

right
Solar reflections
The 11-story-high McMath solar telescope at Kitt Peak National Observatory in Arizona. Sunlight captured by a mirror (heliostat) on top is reflected down the inclined tunnel deep underground, then back up to an observation room, where an image is formed.

left
Skylab's Sun
A huge solar prominence leaps high above the Sun's limb in a dramatic false-color image snapped on board Skylab in 1973. Three teams of three astronauts spent an accumulated time of nearly six months on the space station in 1973/74, with the Sun as the major target for astrophysical investigations.

Probing the Planets

For most of the Space Age, scientists have launched astronomy satellites to study the distant stars. To study the planets and other members of the solar system, they launch different kinds of spacecraft—space probes. Unlike satellites, probes are designed to escape completely from Earth's powerful gravity. To do so, they must be launched at what is called the escape velocity, which on Earth is a speed in excess of some 25,000 miles per hour (40,000 km/h).

Although it missed its target planet, Venus, by thousands of miles, the probe Mariner 2 made the first deep-space discovery in 1962. It revealed that the planet had a heavy carbon dioxide atmosphere. Three years later, Mariner 4 returned the first close-up pictures of a planet, showing the cratered landscape of Mars.

As the years passed, the cameras of these tiny robot explorers revealed the mysteries of one planet after another— Venus, Jupiter, Saturn, Uranus and Neptune. To date, Pluto, in the remote outer fringes of the solar system, is the only planet not yet visited by a space probe. However, it is likely that one will go there before 2006.

The planets are not the only targets of space probes. Giotto and other spacecraft carefully followed Halley's Comet in 1986. Galileo returned the first images of an asteroid (Gaspra) in 1991. NEAR-Shoemaker landed on the asteroid Eros in 2000, in an amazing feat of interplanetary navigation (see page 128).

left
Mariner 4
The best image frame sent back by Mariner 4 when it encountered Mars in July 1965. It was the first time any planet had been pictured from close quarters.

below
Viking invasion
A Viking probe snapped this fine picture of Mars as it was approaching the planet in 1976. It shows three prominent features: clouds clinging to the volcano Ascreus Mons (left), Mars's "Grand Canyon" (middle) and a frosty Argyre Basin (right).

left
Icy Europa
Cracks and ridges criss-cross the ice that covers Jupiter's moon Europa in this false-color Galileo image. The brown lines show where subsurface material has welled up and stained the ice. The blue area is a smoother icy plain.

right
Descent to Titan
It is October 2004. The Huygens probe is descending by parachute to the surface of Titan. Its carrier craft Cassini is now in orbit around Saturn for a long-term study of its atmosphere, rings and moons.

left
Lift-off!
A giant Titan-Centaur rocket lifts off the launch pad at Cape Canaveral, Florida, on August 20, 1977, carrying the Voyager 2 probe. This was the beginning of an amazing voyage of discovery that still hasn't ended.

right
Seeing red
Both Voyager probes studied Jupiter's Great Red Spot (GRS) closely. The furious eddy patterns around the GRS changed constantly as the wind belts coursed round the planet at high speeds.

left
Voyager's Saturn
A striking image of Saturn returned by Voyager 2 in the summer of 1981. This image is rendered in false color to bring out the greatest detail of the planet's atmosphere. It reveals distinctive cloud bands and some storm centers.

left
Sounds of Earth
The record disk each Voyager probe is carrying to inform any beings that may one day encounter them what we and our Earth are like. They carry a multitude of sounds and images of the natural and man-made world, together with messages of hope, peace and goodwill from world leaders.

Star Voyager

One probe stands out above all others for its outstanding achievements. It is Voyager 2, the first of an identical pair launched in 1977. The prime targets of both were Jupiter and Saturn, and both probes sent back incredible images of the giant planets, their moons and their rings. While Voyager 1 disappeared into interplanetary space, Voyager 2 continued to explore other worlds.

The Voyager mission team had decided to attempt a "grand tour" of the outer planets (except Pluto), taking advantage of a planetary configuration that would not occur again for some 176 years. To achieve this in a reasonable time frame, they placed Voyager 2 in a very precise trajectory so that it could pick up gravitational energy from each planet it visited.

This technique, called gravity-assist, had been successfully employed several times before. It does two things: accelerates a spacecraft and flings it out in a different direction, like a slingshot. After gravity-assist at both Jupiter (1979) and Saturn (1981), Voyager 2 went on to encounter Uranus (1986) and Neptune (1989).

Grande finale

Voyager 2 swooped down to just 3,050 miles (4,900 km) above Neptune's cloud tops. It was an utterly amazing close encounter, considering the probe had been journeying through space for more than 12 years and had traveled 4.4 billion miles (7,000,000,000 km). In its "grand finale," Voyager sent back stunning images of Neptune's largest moon Triton, showing an unexpectedly active surface, with strange, icy geysers. Speaking for the imaging team at Jet Propulsion Laboratory, as Voyager headed for the stars, Laurence Soderblom exclaimed: "Wow! What a way to leave the solar system!"

The Hubble Space Telescope

Among optical telescopes, the Hubble Space Telescope reigns supreme. It can see farther into the universe than any others, and with greater clarity, from its vantage point hundreds of miles above Earth's obscuring atmosphere. It is named for the US astronomer Edwin Hubble, who pioneered detailed study of the galaxies during the first quarter of the twentieth century using the famous 100-inch (2.5-m) Hooker telescope at Mount Wilson Observatory. The HST's mirror is comparable in size, with a diameter of 95 inches (2.4 m). But it is a much more effective light-gatherer, not only because it operates in space, but also because of its supersensitive CCDs and "state of the art" electronics.

Work began on the HST in the late 1970s, but it was not launched until 1990. "First light" revealed image-blurring due to a flawed primary mirror, and the whole project seemed doomed. But a daring recovery-and-repair mission three years later sharpened the HST's vision and imagery to design specifications. Updated systems and instruments installed on subsequent servicing missions have made the HST an astronomical observatory without equal in the study of the universe.

He might have been a world-class heavyweight boxer; he might have been a brilliant lawyer. But instead Edwin Powell Hubble (1889–1953) chose to to become an astronomer.

Born in 1889 in Marshfield, Missouri, Hubble later moved with his family to Chicago, where he gained a degree in mathematics and astronomy at the University. By 1910, he was studying at Oxford University, gaining a bachelor's degree in Jurisprudence two years later. In 1913, back in the US, he opened a law practice in Kentucky.

Investigating Faint Nebulas

However, the lure of the heavens proved too strong for Hubble, and he returned to Chicago to become a graduate student at Yerkes Observatory. In 1917, he gained a doctorate with his thesis "Photographic Investigations of Faint Nebulae," in which he suggested that spiral nebulas might lie outside our own star system.

After a stint with the American Expeditionary Force in France during World War I, Hubble accepted an invitation by George Ellery Hale to work at Mount Wilson Observatory. It was there, in 1919, that he began observing with the newly completed 100-inch (2.5-m) Hooker Telescope, then the biggest in the world. The Hooker still wasn't powerful enough to resolve individual stars in spiral nebulas, although it could detect pinpricks of light in them that some thought might be novas.

In 1923, Hubble spotted one in a photograph of the Great Spiral in Andromeda. He then checked previous photographic plates and found points of light in the same position but with different brightnesses. He plotted the light curve of the source and found that it had the typical characteristics of a Cepheid variable. Applying Henrietta Leavitt's period-luminosity law, Hubble deduced that the Cepheid was an astonishing 900,000 light-years away.

This meant that the Andromeda Nebula was way beyond the confines of our own Milky Way Galaxy. It had to be a separate star system—a separate galaxy.

At the time the Milky Way was estimated to be 300,000 light-years across—three times its actual size. Hubble's value of the distance to the Great Spiral on the other hand was a gross underestimate; the true distance is about 2.5 million light-years.

Hubble's Tuning Fork

Hubble eventually prepared a paper announcing this momentous discovery which Henry Norris Russell presented on New Year's Day 1925. Shortly after, Hubble introduced a system for classifying the galaxies, or "extragalactic nebulae," as he preferred to call them. There were irregulars, with no particular shape, and regulars, with a characteristic shape. He devised the famous "tuning fork" system, still used, which classifies the regulars by their shape into ellipticals, spirals and barred spirals (see page 70).

left
Edwin Hubble
The foremost US astronomer of modern times, whose work led to an appreciation of the true scale of the universe.

Hubble also continued, by looking for Cepheids, to determine the distance to other galaxies. Meanwhile, at Lowell Observatory in Arizona, astronomer Vesto Slipher was estimating the speeds at which galaxies were moving by examining the spectrum of their light. He found that almost all had a red shift in spectral lines, indicating that they were rushing away from us, and at unheard-of speeds.

An Expanding Universe

By correlating his distance values with Slipher's speeds of recession, Hubble made an astonishing discovery. The farther away a galaxy was, the faster it receded. Hubble announced this startling fact in 1929, and it became known as Hubble's Law. It led to the fundamental concept of modern cosmology — that the universe is expanding.

Just as happened in World War I, World War II interrupted Hubble's work in astronomy. He became chief of ballistics and director of the Supersonic Wind Tunnel Laboratory in Maryland. After the war, Hubble became involved with the 200-inch (5-m) Hale Telescope project at Mount Palomar Observatory. When the telescope became operational in 1949, it was fitting that Hubble should begin observations. With this powerful instrument, he could at last resolve the stars in the Great Spiral and other spirals.

By this time, Hubble had only a few years to live. He had developed a heart condition and in 1953 died from a cerebral thrombosis. It is altogether appropriate that the Hubble Space Telescope, which is every day expanding our knowledge of the universe, is named for the man who proved that the universe itself is expanding.

above
Hubble sunrise
Sunlight turns the HST into liquid gold as the Sun peeps over the limb of the Earth, hundreds of miles away. It was one of the 16 sunrises per day the STS-109 mission astronauts experienced on their servicing mission in March 2002.

Getting Into Orbit

In 1923, the same year that Edwin Hubble proved that spiral nebulas were external galaxies, German rocket scientist Hermann Oberth wrote a groundbreaking book about space travel called (in translation), *The Rocket into Interplanetary Space*. In it he discussed the possibility of a space telescope that could be attached to an orbiting space station, or even mounted on a small asteroid for stability.

Oberth's ideas were far ahead of his time. Not until 1946 did the concept of a space telescope resurface, in a paper written by the US astronomer Lyman Spitzer. It was entitled "Astronomical Advantages of an Extra-Terrestrial Observatory." Only 11 years later, Sputnik 1 went into orbit and space flight became a reality.

The Large Space Telescope
In the 1960s, Spitzer began to lobby NASA and the US Congress to provide funding for a large orbiting telescope. By the decade's end, NASA's first orbiting telescope (OAO-2) was operational, and a committee of the National Academy of Sciences had published a report on "Scientific Uses of the Large Space Telescope." Funding for a proposed 120-inch (3-m) instrument was elusive, and so plans were drawn up for a cheaper 95-inch (2.4-m) version.

Congress eventually approved funding for the smaller instrument, and in 1977 NASA began work on the project, known as the Large Space Telescope (LST). By then, NASA had acquired a partner, the European Space Agency (ESA). Under their agreement, ESA would provide 15 percent of the hardware, in return for 15 percent of observing time. The launch was scheduled for 1983.

Lift-off
By 1979, the detailed design and specifications of the LST had been finalized, and construction began. With the involvement of more than 20 major contractors, a university and three NASA centers, building the LST was far from straightforward. It also proved much more expensive that anticipated.

In the event, the telescope was nowhere near ready for launch in 1983, when it was officially named the Edwin P. Hubble Space Telescope. A possible 1986 launch was delayed by the grounding of the shuttle fleet following the Challenger disaster. Even when the shuttle fleet became operational again in 1988, the HST had to wait in line. Not until April 24, 1990, did the telescope ascend into the heavens. Announced Kennedy Space Center's launch commentator: "Lift-off of the space shuttle Discovery with the Hubble Space Telescope — our window on the universe."

How the HST Works
The HST orbits the Earth about once every 95 minutes, in a nearly circular path about 350 miles (560 km) high. This height is well within reach of the shuttles that service it. However, this is a relatively low altitude for a satellite, and its orbit gradually decays because of faint traces of atmospheric gases still present. So, after a service mission, visiting shuttles redeploy it in a slightly higher orbit.

The HST is a big satellite, over 43 feet (13 m) long, with a diameter of 14 feet (4.3 m). It weighs about 12.5 tons (11.5 tonnes). Designed for regular servicing while in orbit, it is outfitted with hand-holds and also grapple fixtures so that it can be gripped by the shuttle orbiter's RMS (robot) arm. Its instruments are modular, so that each can be repaired or replaced without affecting the others.

Two large solar arrays, made up of thousands of solar cells, provide the HST with electricity to keep its batteries charged. The latest panels can supply more than five kilowatts. Unlike the original panels, which were flexible and unfurled, the present ones are rigid. They are also smaller, measuring about 23 feet (7.1 m) long and 8.5 feet (2.6 m) wide.

above
The HST in orbit
The main body tube houses a fairly conventional reflecting telescope, which feeds focused light into instruments in the aft assembly. Data gathered by the telescope is transmitted back to the ground by radio from a pair of antennas. Twin panels (arrays) of solar cells provide the necessary electrical power.

left
Hands-on
STS-82 astronauts Steven Smith (left) and Mark Lee gather their tools in preparation for another day's work on a routine servicing mission. The HST was the first major satellite designed specifically for in-orbit servicing.

Radio antenna

Electronics box

Aperture door (Sunshade)

Light shield

Secondary mirror

Solar cells

Primary mirror

Sun sensor

Solar array

Axial instrument modules

above
Anatomy
A cutaway diagram of the HST, showing essential features. A system of gyroscopes keeps the telescope steady once it has locked onto a target. Instruments and optics combine to make the HST so light-sensitive that it could spot a firefly 10,000 miles (16,000 km) away.

Fine guidance sensors

Aft shroud

Radio antenna

The Telescope Assembly

The optical design of the HST literally mirrors that of Earth-based telescopes. It is a reflecting telescope that uses mirrors to gather and focus light.

The telescope is a Richey-Chretien variant of a standard Cassegrain design. In a Cassegrain telescope, light gathered by the large primary mirror is reflected up to a smaller secondary mirror. Then the secondary reflects the light back down the telescope tube and through a hole in the center of the primary, bringing it to a focus underneath. In the case of the HST, the primary mirror is 95 inches (2.4 m) in diameter.

The Science Instruments
Light focused by the optical system feeds into a variety of cameras and other science instruments, located behind the primary mirror in the aft shroud. The five original instruments on the HST were the High-Speed Photometer, the Wide-Field and Planetary Camera, the Faint Object Camera, the Goddard High-Resolution Spectrograph and the Faint Object Spectrograph. All these instruments were replaced during the first four servicing missions.

Wide-Field and Planetary Camera (WF/PC)

The WF/PC is the most widely used of the instruments on the HST. It comprises two cameras, each using CCDs (charge-coupled devices) to record images. In wide-field mode, it looks at large areas of sky, while the planetary camera provides higher resolution and magnification. The original instrument, WF/PC-1, was removed on the first servicing mission and replaced by WF/PC-2, which contains a system for correcting the flawed optics of the primary mirror. When data from all four CCDs in the instrument are displayed together, it produces a chevron-shaped edge, because the scale of the planetary camera image needs to be reduced to that of the wide-field camera.

Faint Object Camera (FOC)

As its name implies, the FOC was designed to see faint objects in the greatest detail. It used an electronic light detector called an image intensifier. It was so sensitive that, to look at anything brighter than magnitude 21, filters were needed to dim the light, to prevent the detectors from becoming saturated.

The FOC was one of the instruments that COSTAR brought back to full operating capability. After impeccable service, it was replaced on the STS-109 servicing mission in March 2002, by the Advanced Camera for Surveys (see page 176).

High-Speed Photometer (HSP)

The HSP was designed to measure high-speed fluctuations of light from sources of high energy. It was affected by the flawed optics of the primary mirror and the vibrations caused by one of the solar panels. It was removed on the first servicing mission in December 1993 and replaced with the COSTAR assembly, which helped correct the HST's vision.

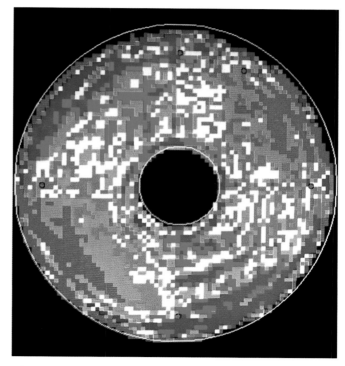

left
Fine limits
Computer scans were used to verify the accuracy of the HST's mirrors, enabling them to be ground and polished to fine limits.

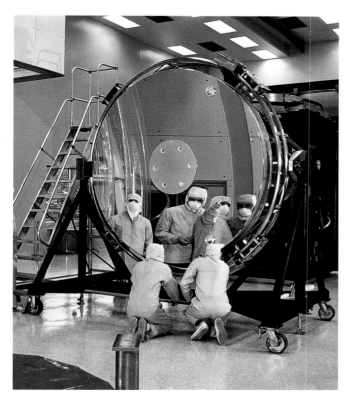

Spectrographs

Spectrographs are designed to produce spectra—smudges of light spread out into separate wavelengths. The dark lines in a spectrum tell astronomers a great deal about the star (or galaxy) that produced the light. Different elements in the star produce different sets of dark lines. A shift in the lines from their normal position indicates that the star may be moving toward us (blue shift) or away from us (red shift), and so on.

The original Goddard High Resolution Spectrograph (GHRS) and Faint Object Spectrograph (FOS) did excellent work after the COSTAR repair. Both were designed to work in the ultraviolet. On the second servicing mission, in February 1997, they were replaced by the Space Telescope Imaging Spectrograph (STIS), which spans visible and some infrared wavelengths as well as ultraviolet rays.

Fine Guidance Sensors (FGSs)

The HST can only function properly if it is directed to a particular part of the heavens and then kept absolutely still. It therefore has three FGSs, which lock onto particular guide stars and can keep the telescope absolutely steady for long periods.

The FGSs also function as the HST's sixth science instrument, for they are able to provide star positions with at least 10 times the accuracy astronomers achieve on the ground. Measuring star positions is called astrometry.

left
Mirror, mirror
Masked engineers examine the completed primary mirror of the HST in a dust-free clean-room. The mirror is made from a special silica/titanium oxide glass with minimal thermal expansion and is one foot (30 cm) thick. It took two years to polish.

HST Images

Data flow
The circuitous route by which HST data reach the user. Observing time on the HST is at a premium, with on average less than half of interested applicants being successful. But some time is still found for amateur astronomers.

H e might have been a world-class heavyweight boxer; he might have been a brilliant lawyer. But instead Edwin Powell Hubble (1889–1953) chose to to become an astronomer.

Born in 1889 in Marshfield, Missouri, Hubble later moved with his family to Chicago, where he gained a degree in mathematics and astronomy at the University. By 1910, he was studying at Oxford University, gaining a bachelor's degree in Jurisprudence two years later. In 1913, back in the US, he opened a law practice in Kentucky.

Investigating Faint Nebulas

However, the lure of the heavens proved too strong for Hubble, and he returned to Chicago to become a graduate student at Yerkes Observatory. In 1917, he gained a doctorate with his thesis "Photographic Investigations of Faint Nebulae," in which he suggested that spiral nebulas might lie outside our own star system.

After a stint with the American Expeditionary Force in France during World War I, Hubble accepted an invitation by George Ellery Hale to work at Mount Wilson Observatory. It was there, in 1919, that he began observing with the newly completed 100-inch (2.5-m) Hooker Telescope, then the biggest in the world. The Hooker still wasn't powerful enough to resolve individual stars in spiral nebulas, although it could detect pinpricks of light in them that some thought might be novas.

In 1923, Hubble spotted one in a photograph of the Great Spiral in Andromeda. He then checked previous photographic plates and found points of light in the same position but with different brightnesses. He plotted the light curve of the source and found that it had the typical characteristics of a Cepheid variable. Applying Henrietta Leavitt's period-luminosity law, Hubble deduced that the Cepheid was an astonishing 900,000 light-years away.

This meant that the Andromeda Nebula was way beyond the confines of our own Milky Way Galaxy. It had to be a separate star system—a separate galaxy.

At the time the Milky Way was estimated to be 300,000 light-years across—three times its actual size. Hubble's value of the distance to the Great Spiral on the other hand was a gross underestimate; the true distance is about 2.5 million light-years.

WF/PC imaging
HST wide-field/ planetary-camera (WF/PC) views of a barred-spiral galaxy (NGC 1808), a "starburst" galaxy where vigorous star formation is taking place. In the left-hand frame, a combined WF/PC image is superimposed over a ground-based telescope picture. The smaller square shows the planetary camera view shrunk down to the same scale as the wide-field view. This is what causes the chevron edge in many HST images. The more detailed magnified planetary-camera image is shown on the right.

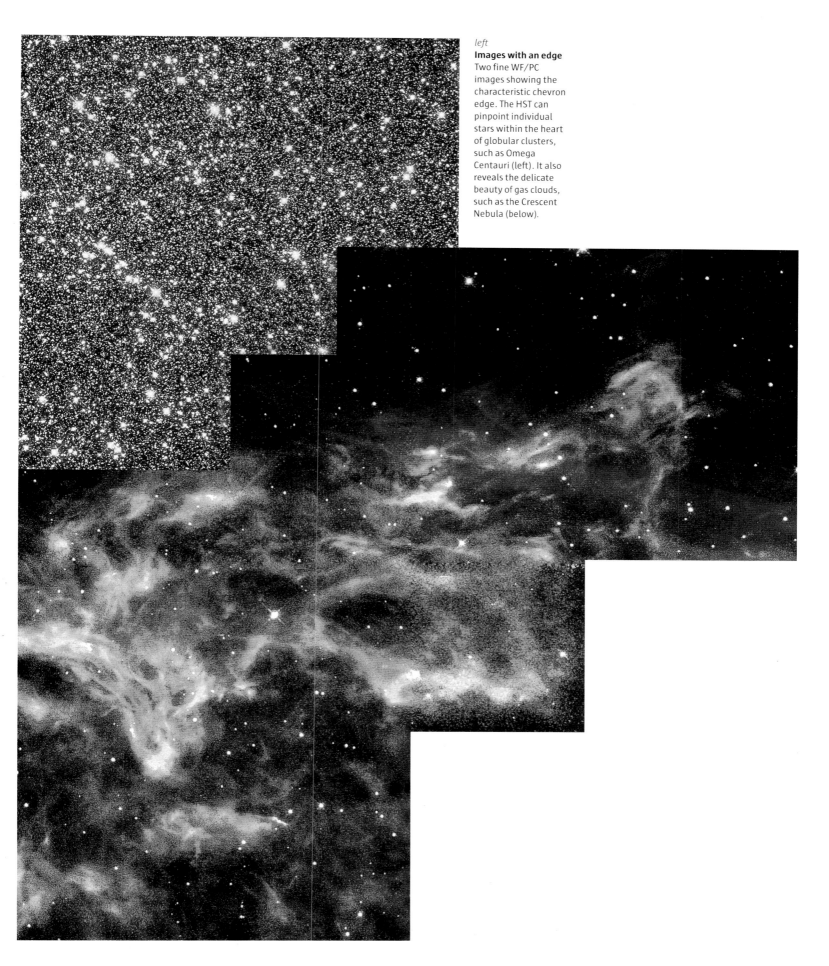

left

Images with an edge
Two fine WF/PC
images showing the
characteristic chevron
edge. The HST can
pinpoint individual
stars within the heart
of globular clusters,
such as Omega
Centauri (left). It also
reveals the delicate
beauty of gas clouds,
such as the Crescent
Nebula (below).

Colorful conclusion
Delicate filaments of glowing gas in Cassiopeia mark the remains of a star that blew itself to bits in a supernova explosion that astronomers witnessed in 1572. The HST team has combined and colored images taken through the filters to highlight the presence of different chemical elements. For example, deep blue shows regions rich in oxygen; red, regions rich in sulfur.

above

Seeing the invisible
These two different views of the same region of the Cone Nebula show how different wavelengths tease out additional details. At left is a visible-light image from the Advanced Camera for Surveys; at right, an infrared image from NICMOS. Note that infrared light has penetrated the dusty haze to reveal a clutch of hot stars.

left

Remote cepheids
This remote galaxy is NGC 4603, one of the cluster of galaxies in Centaurus. The HST has been able to pinpoint Cepheid variables in the galaxy, establishing its distance at 108 million light-years.

A Flawed Debut

Seven years after its original planned launch date, the HST arrived at the Kennedy Space Center in October 1989. It was installed in the payload bay of shuttle orbiter Discovery on March 29, 1990, with launch planned for April 10. The HST launch mission was designated STS-31.

However, problems with the orbiter's APUs (auxiliary power units) caused the launch to be scrubbed at the T-4 minutes hold in the countdown. Two weeks later, on April 24, the countdown went all the way, and Discovery and the HST thundered into the heavens.

Next day, with Steve Hawley at the controls, Discovery's robot arm picked up the HST and lifted it out of the payload bay. When Hawley sent commands to unfurl the twin solar panels, one unfurled but the other jammed. Inside Discovery's airlock, astronauts Kathy Sullivan and Bruce McCandless were suited up, ready to go on EVA to free the jammed solar panel manually. But eventually it freed itself.

Hawley released the robot arm's hold on the HST, which stayed close by. Two days later, it took two attempts to open the aperture door of the HST. It was at this point that Story Musgrave, the CapCom (capsule communicator) at Mission Control, radioed: "Discovery, Hubble is open for business."

First Light

As it happened, the HST wasn't quite "open." The first few weeks after launch were used for equipment, systems and engineering checks. Not until May 20, 1990, did the HST experience "first light" — this is the time when a telescope becomes fully operational and returns the first image.

This long-anticipated event unfolded before the eyes of journalists from around the world at the Space Telescope Science Institute, who cheered as an image flashed onto the screen before them. The portrait of the open star cluster NGC 3532 was actually a somewhat boring image, but it was greeted with elation by the press and with more than a little relief by mission scientists. This was the first evidence that the HST actually worked!

Behind the scenes, however, HST scientists were worried. They realized that the first-light image was blurred. And no matter what they did, they couldn't make it sharp. The HST wasn't focusing properly! Word soon leaked to the once adulatory press, who were quick to pour scorn: "Pix nixed as Hubble sees double" screamed one memorable headline.

The HST team soon realized that their US$1.5 billion brainchild had flawed optics. The primary light-gathering mirror was suffering from a classic curved-mirror defect known as spherical aberration. The mirror had an incorrect curvature, which caused light rays from different parts of the mirror to be brought to a focus at different points.

It turned out that the primary mirror deviated from the correct curvature by a miniscule amount — just two microns, or about one-fiftieth of the width of a human hair. Still, this was enough to blur the images. The defect had apparently been caused in manufacturing and had not been detected during subsequent tests.

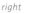

left
Smile please
Discovery's crew poses for the traditional on-board photograph. The astronauts are Charles Bolden (at top), then, from left to right: Loren Shriver, Kathryn Sullivan, Bruce McCandless and Steven Hawley.

right
Into orbit
On April 25, 1990, Discovery's robot arm plucks the HST from the payload bay and places it in orbit. It is ready for business, or so everyone thought.

far left
Lift-off!
All engines blazing, shuttle orbiter Discovery thunders away from the launch pad at the Kennedy Space Center carrying the HST. It is April 24, 1990. The HST is late — very late.

left
Stormy giant
In the fall of 1990, astronomers reported seeing a white spot on Saturn, marking the site of an exceptional storm. This is one of the time-lapse images the HST took to track how the storm developed; it has been color-enhanced.

Emergency Service

left
Capture
It is December 4, 1993. Shuttle orbiter Endeavour maneuvers closer and closer to the ailing HST, seen here against a backdrop of the Indian Ocean off Australia's west coast. Soon the HST will be safely berthed in Endeavour's payload bay.

left
New for old
On their EVA of December 7, Jeff Hoffman and Story Musgrave install the new WF/PC-2. Here Hoffman is seen with the old instrument, which he will stow in the payload bay.

So here was a telescope hundreds of miles away in space that was returning blurred pictures due to faulty optics. While NASA and the Space Telescope Science Institute debated what to do, computer experts were devising a way to manipulate the incoming data in order to improve the quality of the images. This computerized image-processing technique, called deconvolution, involved determining where light was missing from the image and restoring it.

Deconvolution and longer exposure times certainly helped to produce acceptable images, but they were still only on a par with the best images taken by the most powerful observatories on Earth under good conditions. The mammoth leap in observational astronomy that scientists were expecting from the HST had not materialized.

How to fix the HST—that was the problem. Some suggested that it should be recovered, returned to Earth, refurbished here and then redeployed. But the consensus was that the HST should be repaired in orbit. After all, it been designed for in-orbit servicing.

COSTAR to the Rescue

The Hubble team soon devised a means of correcting the telescope's defects. They would use an ingenious piece of equipment with 10 small mirrors, called COSTAR (corrective optics space telescope axial replacement). COSTAR would clarify the Hubble's vision rather like eyeglasses cure defective eyesight. Its mirrors would refocus the light coming from the defective primary and feed it to the Faint Object Camera (FOC) and spectrographs. However, in order to make room for COSTAR aboard the HST, the High-Speed Photometer (HSP) would have to be removed.

The HST team would also replace the existing Wide Field and Planetary Camera (WF/PC-1) with a "clone" (WF/PC-2) that had reconfigured mirrors. These, too, would combat the effects of the mirror's aberration.

Mission Accomplished

By December 1993, NASA was ready to launch a rescue and repair mission, designated STS-61. And just in time. By now, the HST really was in trouble and on the verge of becoming useless. Three of its six gyroscopes had failed, and so had two of its memory banks. In addition, its solar panels vibrated every time it passed between day and night or light and shade, which happened 16 times every 24 hours.

Just before dawn on December 2, 1993, shuttle orbiter Endeavour blasted off on the vital first servicing mission,. It rendezvoused with the HST two days later. Using the robot arm, Endeavour captured the telescope and berthed it in the payload bay. On December 5, veteran spacewalkers Story Musgrave and Jeff Hoffman made the first of five EVAs (extravehicular activities) to make repairs. It took them nearly eight hours to replace the faulty gyroscopes and other electronic parts.

On December 6, Kathryn Thornton and Tom Akers went on EVA to replace the two solar panels. Musgrave and Hoffman returned the next day to install the new WF/PC. On December 8, Thornton and Akers removed the High-Speed Photometer (HSP) and fitted COSTAR into place. Musgrave and Hoffmann carried out the last of the five spacewalks later the same day, bringing the total EVA time for the mission to more than 35 hours—a shuttle record. On December 9, Endeavour's robot arm lifted the HST out of the payload bay and back into independent orbit. Endeavour itself returned to base four days later.

Second First Light

From NASA's point of view, the STS-61 servicing mission had gone amazingly well. But it remained to be seen whether the repair team's "eye surgery on the patient" had corrected the telescope's faulty vision. On December 18, 1993, the HST science team gathered to witness first light of the refurbished instrument. An image of a star, Melnick 34 in the 30 Doradus (Tarantula) Nebula, appeared on the screen as a bright point of light—not spread out or blurred. The repair mission had worked.

By January 13, 1994, the HST team had acquired a portfolio of spectacular images taken with the rejuvenated telescope, which they shared with the press. "The patient," an HST spokesman said, "has a new vision of incredible clarity."

New Vision

Over the next four years, the HST returned spectacular images by the hundreds, proving that it had indeed opened up a new

above
Nearly done
On the final EVA of the STS-61 repair mission, Story Musgrave hitches a ride on the robot arm to fit protective covers over the magnetometers, instruments designed to detect magnetic fields.

left
New vision
Two HST images that tell the story of the success of the first make-or-break service mission. The image of the spiral galaxy M100 before the mission (left) is little better than views from ground telescopes. The image of the galaxy after the repair is infinitely clearer.

window on the universe. In February 1997, a planned second servicing mission (STS-82) was underway. Discovery lifted off with a number of new instruments to improve the telescope's visual acuity and range.

Spacewalking astronauts, working alternately in pairs — first Mark Lee and Steve Smith and then Greg Harbaugh and Joe Tanner — carried out five EVAs to install the new equipment. On the first and most important EVA, Lee and Smith removed the two existing spectrographs and replaced them with the Space Telescope Imaging Spectrograph (STIS) and the Near-Infrared Camera and Multi-Object Spectrometer (NICMOS). The STIS would provide improved spectral resolution, while NICMOS would extend the sight of the HST into near-infrared wavelengths.

Subsequent EVAs saw the replacement of a number of instruments and equipment, including one of the Fine Guidance Sensors (FGSs), a data recorder and a Reaction Wheel Assembly (RWA). On the final EVA, astronauts Lee and Smith repaired damaged thermal insulation on the outside of the HST, which protects the internal structure and instruments from fluctuations in temperature.

Into Hibernation

A third HST servicing mission was planned for June 2000, but one after another, its attitude-maintaining gyroscopes began to fail in 1997, 1998 and 1999. So the HST team decided to split the objectives of the mission in two, with the first launch in December 1999. On November 17 the fourth gyro failed, leaving just two gyros functioning — not enough to position the instrument. The telescope had to be shut down — it was put into "hibernation mode."

Discovery blasted off on the STS-103 service mission (designated SM3A) on December 19, 1999. On three spacewalks, astronauts replaced the faulty gyros, fitted kits to improve the telescope's batteries and installed a new computer that was 20 times faster and had six times as much memory as the original.

Advanced Imaging

Shuttle orbiter Columbia sped into orbit on March 1, 2002, on the STS-109 service mission (designated SM3B). The crew kept a rigorous schedule in order to install all of the new and upgraded instruments and equipment.

The critical new instrument was the Advanced Camera for Surveys (ACS), which replaced the Faint Object Camera (FOC). The FOC had been "state of the art" in the 1980s, but digital imaging had improved dramatically since then.

The ACS comprises three separate instruments: the High-Resolution Channel, the Solar-Blind Channel and the Wide-Field Channel. The High-Resolution Channel takes detailed

left
Instrument update
Astronauts move in to install new equipment in the HST on the STS-82 mission in February 1997. Joe Tanner stands on the platform on the robot arm, with Greg Harbaugh nearby. On five EVAs during the mission, astronauts notched up a total time of more than 33 hours.

right
The Pistol
The newly installed NICMOS instrument gave the HST the ability to see in the near-infrared. It was soon returning spectacular images — which of course had to reproduced in false-color. This image centers on the Pistol Star and the surrounding nebula. It is one of the most luminous stars we know.

images of the inner regions of galaxies and searches for extrasolar planets. The Solar-Blind Channel, which blocks visible light to enhance ultraviolet sensitivity, studies planetary weather in our own solar system. The Wide-Field Channel helps astronomers study the nature and distribution of the galaxies on a broad scale.

The ACS was installed on the fourth of the five EVAs of the mission. Earlier, astronauts had replaced the solar arrays with a smaller but more powerful set and installed a new Reaction Wheel Assembly (RWA).

On the fifth EVA, a new neon cryocooler—a refrigeration unit—was fitted for the NICMOS. Working in the infrared, its instruments must be cooled to around minus 330 degrees Fahrenheit (–200°C). The original nitrogen-ice coolant had run out in January 1999, earlier than expected. Installation of the new cooler would bring the instrument back into operation.

The five EVAs on STS-109 broke the duration record for a single flight, totaling nearly 36 hours. Over the four HST servicing missions to date, 14 astronauts had spent a total of more than 129 hours on maintenance work.

Toward the Future

On another planned servicing mission (SM4), scheduled for late 2003 or 2004, a third-generation WF/PC will bring greater sensitivity to this most widely used HST instrument. A device called the Cosmic Origins Spectrograph will replace the COSTAR assembly. Barring unforeseen circumstances, these instruments should serve the HST until the end of its planned life in 2010.

By then the HST's successor should be ready for launch. Originally known as the Next Generation Space Telescope, it has been renamed the James Webb Space Telescope (JWST) for James Webb, NASA Administrator during the Apollo era. This is a departure from NASA's tradition of naming astronomy spacecraft after scientists (such as Hubble).

The JWST will be quite a different instrument from the HST. Also, it will not be launched into orbit around the Earth, but into a gravitationally stable location in space known as Lagrangian Point 2, about 930,000 miles (1,500,000 km) away.

The specifications for the JWST are far from finalized, but it is expected to have a primary mirror at least 20 feet (6 m) across, some two-and-a-half times the size of the HST's. This much larger mirror should enable the JWST to detect objects 400 times fainter than the HST. It will be made up of a number of hexagonal segments, which will be folded for launch and then put together in orbit. A sunshield as big as a tennis court will also deploy after launch to protect the spacecraft from the Sun's heat and light.

Revealing the Dark Ages

The JWST will also view the universe at different wavelengths from the HST. The HST observes over a wide range of wavelengths, from the ultraviolet to the near-infrared. The JWST will make observations primarily in the infrared. Studies of young stars and planetary systems are best performed at infrared wavelengths. So are studies of remote galaxies, whose light has been severely red shifted because of their speed of recession.

Astronomers hope that the JWST will shed much more light on what are called the "dark ages of the universe," the period of time when the first stars and galaxies were forming. How this new space telescope will shape up remains to be seen, but what is clear is that the HST is going to be a very hard act to follow.

below
New wings
With smaller but more powerful solar arrays, NICMOS back in working order, and a new and more advanced camera, the HST begins to orbit independently at the end of the STS-109 mission.

following pages
Hubble's heritage
The Hubble Space Telescope will long be remembered for stunning images like this, of the Little Ghost Nebula. The nebula glows in Technicolor brilliance as it is excited by the intense ultraviolet radiation emitted by the dying star in the center.

Glossary of Terms

absolute magnitude The true brightness of a star; the magnitude that would be observed from a distance of 10 parsecs (33 light-years)

accretion disk A disk of gas and dust that forms around newborn stars and black holes

active galaxy A galaxy with an exceptionally high energy output (often as radio waves) thought to originate from a massive black hole

active optics A technique for building large telescope mirrors. It uses a number of mirror segments that are aligned together into a perfect shape by computer-controlled actuators

antimatter Matter made up of particles that are the exact opposite of ordinary particles, such as positrons (antielectrons, with a positive electric charge) and antiprotons (with a negative charge)

apparent magnitude The brightness of a star as we see it in the sky

asterism A close grouping of stars in the sky

asteroid A chunk of rock or metal circling in space, mainly in a band (the asteroid belt) roughly mid-way between the orbits of Mars and Jupiter; also called a minor planet

astrology An ancient belief that human lives and fortunes are somehow affected by the relative positions of the Sun and the planets among the stars

astronomical unit (AU) The average distance between the Earth and the Sun — 93,000,000 miles (150,000,000 km)

astronomy The study of the heavens and heavenly bodies

atmosphere The layer of gases that surrounds the Earth and other planets; a unit of pressure, being the pressure exerted by the Earth's atmosphere at sea level

atom The smallest unit of matter, with a central concentration of mass (called the nucleus) circled by electrons

aurora A glow in the atmosphere of a planet's polar regions due to charged particles interacting with molecules in the upper air. On Earth, in the Northern Hemisphere, the light displays are called the aurora borealis, or Northern Lights, and in the Southern Hemisphere the aurora australis, or Southern Lights. The HST has spotted auroras on Jupiter and Saturn

barred-spiral A kind of spiral galaxy that has a band of stars extending through its nucleus

Big Bang The generally accepted theory that the universe was created in a fantastic explosion roughly 12 to 15 billion years ago

Big Crunch The theory that the universe will end with all matter coming together and being crushed to nothing; the opposite of the **Big Bang**

billion One thousand million: 1,000,000,000

binary A double-star system in which the two stars are bound by gravity and orbit one another

black hole A region of space that has such powerful gravity that not even light can escape. It marks the death of a white dwarf (Type I) or a supermassive star (Type II).

blazar An active galaxy with radiation that seems (to us on Earth) to be coming straight from its core

blue shift A shift in the lines in the spectrum of a star toward the blue end, indicating that the star is approaching us

brightness *See* **absolute magnitude**, **apparent magnitude**

brown dwarf A small, failed star, whose nuclear furnace has never lit up. It produces some heat but little light

CCD Charge-coupled device, used in most modern telescopes (including the HST) to acquire images

celestial sphere An imaginary dark sphere that appears to envelop the Earth, on the inside of which the stars are fixed. It appears to revolve round the Earth once a day from east to west

Cepheid A variable star that changes brightness over a regular, short period; used to estimate distances to remote galaxies

chromosphere The colorful inner atmosphere of the Sun that becomes visible during a total solar eclipse

cluster A grouping of stars or galaxies; *see also* **globular cluster, open cluster, supercluster**

coma The bright head of a comet

comet An icy lump that starts to shine when it approaches the Sun

constellation A pattern of stars that seem to be grouped together because they appear in the same part of the sky

core The central region of a planet, Sun or star

corona The outer atmosphere of the Sun, visible from Earth only during a total solar eclipse

cosmic ray A fast-moving, charged atomic particle emanating from space

cosmology The study of the origin and evolution of the universe

cosmos An alternative name for the universe

COSTAR Corrective Optics Space Telescope Axial Replacement; an assembly of mirrors fitted to the HST to correct the flawed vision of its light-gathering primary mirror

crater A circular pit in the surface of a planet or moon, usually made by the impact of a meteorite

crust The solid surface of a planet or moon

dark matter Unseen matter that is thought to make up as much as 90 percent of the mass of the universe

declination Celestial latitude; the distance a star is from the celestial equator, measured in degrees

degenerate matter A kind of highly compressed matter found in white-dwarf and neutron stars

double star A star that looks like a single star but is actually two stars close together. Some doubles are two stars that happen to appear together in our line of sight. Others are pairs of stars that really are close together, called binaries

eclipse When one heavenly body passes in front of another and blocks out its light; *see also* **lunar eclipse, solar eclipse**

eclipsing binary A kind of variable star, which is actually a binary star system. The two stars orbit in our line of sight and periodically eclipse each other. When they do, the overall brightness of the system drops

electromagnetic radiation Waves of energy emitted by stars and other heavenly bodies — gamma rays, X-rays, ultraviolet waves, visible light rays, infrared rays, microwaves and radio waves. They all travel at the same speed (the speed of light) but have different wavelengths

electron A tiny subatomic particle present in all atoms that has a negative electric charge

element A basic building block of matter that cannot be broken down into a simpler chemical substance

elliptical galaxy A galaxy of a round or oval shape that lacks spiral arms

encounter A meeting in space between a spacecraft and a planet or other heavenly body

ESA European Space Agency, the organization that coordinates space activities in Europe

escape velocity The speed at which a body must travel to escape the gravity of another: Earth's escape velociy is about 25,000 miles per hour (40,000 km/h)

EVA Extravehicular activity, work outside a spacecraft; popularly called spacewalking

evening star Either the planet Venus or Mercury, seen in the western sky after sunset

expanding universe The theory that the universe is expanding, evidenced by the recession of the galaxies

extrasolar planet A planet in orbit around a star other than the Sun

extraterrestrial Not of the Earth; an extra-terrestrial being or an alien

falling star A common name for a meteor

false-color image An image that is reproduced in colors that are not those we would see with our eyes

filament A string of galaxy superclusters that extends over a vast distance in space

first light The first time a telescope produces an image

flyby A space mission in which a probe flies past a planet without going into orbit or landing

fusion *See* **nuclear fusion**

galaxy A "star island" in space. Our Galaxy is the Milky Way, often called simply the Galaxy

gamma rays Electromagnetic rays that have the shortest wavelengths and the most energy

gas giant Jupiter, Saturn, Uranus or Neptune, which are all made up mainly of gas and liquid gas and are giant-sized compared with the terrestrial (Earth-like) planets

globular cluster A huge, globe-shaped mass containing hundreds of thousands of stars that are grouped closely together

gravitational lens The distortion of the light from a distant object when it passes through a strong gravitational field, such as that of a galaxy or galaxy cluster. It usually results in multiple images of the object

gravity The force of attraction that exists between bodies. The more massive a body is, the stronger its gravitational attraction

greenhouse effect A situation in which the atmosphere traps heat like a greenhouse and causes global warming

halo The spherical region around a galaxy that contains globular clusters and dark matter

heavens The night sky. Hence the term heavenly bodies: the objects we see in the night sky

helium The second-most common element in the universe, after hydrogen; formed in stars by the nuclear fusion of hydrogen

H-R diagram The Hertzsprung-Russell diagram; a graph that correlates the absolute brightness of stars with their spectral class or temperature

Hubble Constant A quantity describing the speed at which galaxies accelerate away from one another and, therefore, how fast the universe is expanding

Hubble's law The velocity of recession of a galaxy is proportional to its distance. The constant of proportionality is the Hubble Constant

hydrogen The element in the universe that is the most common, the simplest (composed of one proton and one electron) and the lightest

ice world One of many remote icy bodies that exist at the edge of the solar system. Pluto seems to be the biggest one

infrared rays Electromagnetic rays that are slightly longer in wavelength than visible red light; heat radiation

intergalactic Between galaxies

interplanetary Between planets

interstellar Between stars

interstellar matter Matter in the form of gas and dust that is scattered in the space between the stars; sometimes visible as clouds or nebulas

irregular galaxy A galaxy with an irregular shape

Kuiper Belt A region of space at the edge of the solar system that contains millions of icy, comet-like bodies

lava Molten rock spewed out by volcanoes

light Electromagnetic radiation at wavelengths our eyes can detect

light-year The distance light travels in a year — about 6 trillion miles (9,500,000,000,000 km); used as a unit for measuring distances in space; *See also* **parsec**

LMC The Large Magellanic Cloud

Local Group The small cluster of galaxies to which our own Galaxy belongs. It also includes the Andromeda Galaxy and the Magellanic Clouds

luminosity The absolute magnitude (true brightness) of a star compared with that of the Sun

lunar Relating to the Moon

lunar eclipse An eclipse of the Moon, when it enters the Earth's shadow

Magellanic Clouds *See* **LMC, SMC**

magnetic field The region around a star, galaxy or planet where its magnetism acts

magnetosphere A magnetic bubble surrounding a planet that keeps the solar wind at bay

magnitude The brightness of a star; *See* **absolute magnitude**, **apparent magnitude**

main sequence star A star in the prime of its life, located on a diagonal band in the H-R diagram

mantle The rocky layer of a planet between its crust and core

mare A vast, dusty plain on the Moon, named for the Latin word for "sea"; the plural is maria

mass The amount of matter contained in a physical object

matter The "stuff" the universe is made of

meteor The streak of light that appears in the sky when a meteoroid burns up in the Earth's atmosphere

meteorite A lump of rock or metal that falls to the Earth from outer space; the remains of a large meteoroid

meteoroid A piece of rock or metal traveling through space

Milky Way A pale band of stars visible in the night sky that represents a "slice" through our local Milky Way Galaxy

minor planet Another name for an asteroid

Mira variable A variable star like Mira Ceti that changes brightness over a long period; Mira varies over a period of about 11 months

molecule The smallest part of a substance, made up of a number of similar or different atoms linked together by chemical bonds; *See also* **atom**

molecular cloud An interstellar cloud made up of molecules, such as hydrogen; the birthplace of stars

moon The common name for the natural satellite of a planet; the Moon is Earth's only natural satellite

morning star Either the planet Venus or Mercury, seen in the eastern sky before sunrise

NASA National Aeronautics and Space Administration; the organization that coordinates space activities in the United States

near-earth object (NEO) An asteroid that orbits (sometimes dangerously) close to the Earth

nebula A cloud of gas and dust that exists in the space among the stars

NEO *See* **near-earth object**

neutron A subatomic particle with no electric charge, found in the nucleus of every atom except hydrogen

neutron star A very dense star, made up of neutrons packed tightly together. It is a form of degenerate matter

nova A star that suddenly flares up brightly, named for the Latin word for "new"; *See also* **supernova**

nuclear fusion A nuclear reaction in which atoms of a light element (such as hydrogen) fuse, releasing enormous amounts of energy

nuclear reaction A process of change involving the nuclei of atoms

nucleus The center of an atom or a galaxy; also the small, solid part of a comet; the plural is "nuclei"

observatory The place where astronomers work and make their observations

Oort cloud A huge spherical cloud of icy cometary bodies that more or less marks the edge of the solar system

open cluster A loose grouping of up to a few hundred stars

orbit The path in space one body follows when it circles around another. Most orbits are elliptical (oval) rather than circular

orbital velocity The minimum speed a body must have to remain in orbit around another. For a satellite circling a few hundred miles above the Earth, the orbital velocity is around 17,500 mph (28,000 km/h)

parallax The shift in position of a nearby object against a distant background when viewed from different points; used by astronomers to determine the distances to a few hundred of the nearest stars

parsec A unit used by astronomers to measure distances in space; equal to about 3.3 light-years

period A length of time in a repeating cycle; for example, in a Cepheid, the time between peaks of brightness

period-luminosity law The period of variation of a Cepheid variable is directly related to its absolute magnitude, or luminosity

phases Apparent changes in the shape of the Moon or other heavenly body (like Venus), due to more or less of its surface being lit by the Sun

photon A particle of electromagnetic radiation

photosphere The bright, visible surface of the Sun or other star

planet One of the nine main bodies that circle in space around the Sun, including the Earth. Many other stars have planets too, known as extrasolar planets.

planetary nebula A cloud of gas and dust given off by a dying star with a mass similar to that of the Sun

planetesimal A small body formed early in the life of the solar system. Planetesimals congregated together to form the planets

Population I stars Relatively young, hot stars that are typical of spiral arms and contain an abundance of metals, such as O and B stars, supergiants and Cepheids

Population II stars Relatively old red and yellow stars that are typical of the galactic bulge (nucleus) and globular clusters

probe A spacecraft that travels deep into space to explore heavenly bodies

prominence A fountain-like eruption of incandescent gas on the Sun that often follows the invisible loops of the Sun's magnetic field

proplyd Short for "protoplanetary disk"; a disk of matter around a newborn star from which planets might eventually form

proton A subatomic particle found in the nucleus of all atoms. The simplest element, hydrogen, has a single proton in the nucleus

protostar An early stage in the birth of a star, before nuclear processes have begun

pulsar A rapidly rotating neutron star that flashes pulses of radiation toward us, like a celestial lighthouse

quasar Short for "quasi-stellar"; a kind of active galaxy that emits energy from a small, central region. So named because it appears to be star-like, it is actually much more remote than stars and is as bright as many galaxies

radar The technique of bouncing radio waves off objects used in astronomy to map the surface of a planet

radiant The point in space from which the meteors in a meteor shower appear to originate

radio astronomy The branch of astronomy concerning the study of radio waves from space reaching Earth

radio waves Electromagnetic radiation with the longest wavelengths

red giant A large, red star that is near the end of its life. The Sun will swell up 20 to 30 times to become a red giant in about 5 billion years' time

red shift A shift of the dark lines in the spectrum of a star or galaxy toward the red end, indicating that it is moving away from Earth

reflector A reflecting telescope, which uses mirrors to gather and focus incoming light

refractor A refracting telescope, which uses lenses to gather and focus incoming light

resolving power The ability of a telescope to distinguish fine detail

retrograde motion Motion in the opposite direction from normal. Some of Jupiter's moons have retrograde motion, circling the planet in the opposite direction from the others.

right ascension Celestial longitude, measured in hours of sidereal time–time according to the stars

satellite A small body that travels in orbit around a larger one; a moon. It is also the usual term for an artificial, man-made satellite.

sea A flat, dusty plain on the Moon; See also **mare**

seasons Regular changes in the temperature and weather caused by the tilt of the Earth's axis in space. Other planets have seasons too, most noticeably Mars.

seeing The quality of the image produced in a telescope, which reflects the clarity and steadiness of the atmosphere

SETI The search for extraterrestrial intelligence

Seyfert galaxy An active galaxy with an exceptionally bright center

shepherd moon A moon that orbits close to a planet's rings and appears to keep the ring particles in place

shooting star A common name for a meteor

SMC The Small Magellanic Cloud

solar Relating to the Sun

solar eclipse An eclipse of the Sun in which the Moon partly (partial eclipse) or completely (total eclipse) covers the surface of the Sun and blots out its light; See also **eclipse**

solar flare A powerful explosion near the surface of the Sun that injects high-speed particles into the solar wind

solar system The Sun and the bodies that travel with it through space, primarily the planets, their moons and the asteroids

solar wind A stream of electrified particles given off by the Sun

spectral class The group in which a star is classified according to its spectrum

spectral lines Dark lines in the spectrum of a star or galaxy

spectroscope/spectrograph An instrument that splits light into a spectrum of different wavelengths, displaying the spectral lines

spectrum A band of rainbow colors produced when sunlight or starlight is split into separate wavelengths

speed of light The speed at which light and other electromagnetic radiation travels —around 186,000 miles (300,000 km) per second; it is the fastest speed possible

spiral galaxy A galaxy with curved arms coming out of a central bulge, or nucleus

standard candle An object that astronomers can use to estimate the distance to remote galaxies

star A huge globe of searing hot gases that gives off energy as light, heat and other radiation

stellar Relating to a star or stars

subatomic particle A particle that is smaller than an atom

sunspot A dark patch on the Sun's surface that is cooler than surrounding areas. It varies according to an 11-year cycle, called the solar (or sunspot) cycle

supercluster A grouping of many clusters of galaxies occupying vast regions of space

supergiant The largest type of star, typically hundreds of times wider than the Sun

supernova A catastrophic explosion of a supergiant star (Type II supernova) or— even more powerful—of a white dwarf (Type I supernova)

terrestrial Relating to the Earth

terrestrial planets The planets Earth, Mercury, Venus and Mars, which are made up mainly of rock

trillion One million million: 1,000,000,000,000

true brightness See **absolute magnitude**

ultraviolet rays Electromagnetic rays that have a wavelength just shorter than that of visible violet light

universe Space and everything it contains: galaxies, stars, planets and energy

van Allen belts Donut-shaped regions of intense radiation around the Earth. There are similar regions on other planets

variable star A star that varies in brightness, usually because of an internal process; see also **Cepheid, eclipsing binary, Mira variable**

voids Vast regions of empty space between the superclusters of galaxies that make up the universe

wavelength The distance between the peaks or troughs of a wave motion, such as a wave of electromagnetic radiation

weightlessness The strange condition astronauts experience when traveling in orbit, in which their bodies appear not to have any weight, and gravity seems not to exist. This state is properly termed microgravity

white dwarf A small, very dense star, about the same size as the Earth but with a mass similar to that of the Sun; it marks a late stage in the life of Sun-like star

WIMP Weakly Interactive Massive Particle, thought to be one kind of dark matter

X-rays Penetrating electromagnetic rays with a wavelength longer than gamma rays but shorter than ultraviolet rays

zodiac An imaginary band in the heavens through which the Sun, Moon and planets appear to travel. Hence, the constellations of the zodiac

Landmarks in Astronomy

585 BC
Thales of Miletus correctly predicts the total solar eclipse of this year

350 BC
Aristotle proves that, since the Earth throws a curved shadow on the Moon, it is round and not flat

280 BC
Aristarchus of Samos suggests that the Earth might orbit the Sun, instead of vice versa

240 BC
Eratosthenes measures the size of the Earth with considerable accuracy

The first recorded appearance of what becomes later known as Halley's Comet

140 BC
Hipparchus creates a catalogue of over 1,000 stars and establishes the magnitude system of measuring the brightness of stars, which is still used today

AD 150
Ptolemy of Alexandria compiles the *Almagest*, which sums up ancient astronomical knowledge

813
Al-Mamun founds the Baghdad school of astronomy and translates the *Almagest* into Arabic

1054
Chinese astronomers witness a supernova in Taurus, which gives rise to the supernova remnant we know as the Crab Nebula

1433
Ulugh Beigh establishes the world's finest observatory at Samarqand (in modern-day Uzbekistan)

1543
Nicolaus Copernicus resurrects the concept of a solar system in his seminal work "De Revolutionibus Orbium Coelestium" (Concerning the Revolution of Celestial Spheres)

1572
Tycho Brahe observes a supernova in Cassiopeia

1595
David Fabricius discovers the variability of Mira Ceti

1600
Giordano Bruno is burned at the stake for the heresy of supporting Copernicus's concept of a solar system

1609
Johannes Kepler publishes the first of his laws of planetary motion, which states that planets move in elliptical orbits around the Sun

Galileo builds a telescope and makes the first telescopic observations of the heavens

1666
Isaac Newton formulates his laws of gravity and investigates the visible spectrum

1672
Newton builds the first reflecting telescope

1675
Greenwich Observatory is founded in England

Ole Romer measures the speed of light

1682
Edmond Halley observes the bright comet that is subsequently named after him; he calculates that it is a regular visitor to the Earth's skies and (correctly) predicts its return in 1758

1728
James Bradley introduces the idea of using parallax to measure stellar distances

1781
William Herschel discovers Uranus, the first new planet discovered since ancient times

Charles Messier draws up a list of clusters and nebulas, to distinguish them from comets

1801
Guiseppe Piazzi discovers Ceres — the first and biggest asteroid

1802
William Wollaston observes dark lines in the Sun's spectrum

1815
Joseph Fraunhofer studies in detail the dark lines in the solar spectrum, henceforth known as Fraunhofer lines

1838
Using the parallax principle, Friedrich Bessel makes the first accurate measurement of stellar distance — 65 trillion miles, (105,000,000,000,000 km) to the star 61 Cygni

1843
Heinrich Schwabe discovers the 11-year solar (sunspot) cycle

1845
Lord Rosse completes his giant reflector ("the Leviathan of Parsonstown") at Birr Castle and discovers the spiral nature of some nebulas, including the Whirlpool Galaxy (M51)

1846
Johann Galle discovers an eighth planet — Neptune

1877
Giovanni Schiaparelli reports seeing "canali" (channels) on Mars, which is mistranslated from Italian as "canals," or artificial waterways; this gives rise to the idea of a Martian civilization

1888
Johann Dreyer's *New General Catalog of Nebulae and Clusters* (NGC) is published, which assigns numbers (that are still used) for the identification of clusters, nebulas and galaxies

1894
Percival Lowell founds the Lowell Observatory in Flagstaff, Arizona, specifically to study Mars

1905
Mount Wilson Observatory is founded in Pasadena, California

Ejnar Hertzsprung publishes the first of several papers about classifying stars

1908
The great Siberian "meteorite" lands; it is now thought to have been the nucleus of a comet

1912
Henrietta Leavitt discovers the period-luminosity law for Cepheid variables: that the period of variability of a Cepheid is related to its luminosity, or true brightness

1913
Henry Norris Russell publishes his ideas on stellar evolution, his work dovetailing with that of Hertzsprung; this leads to the famous Hertzsprung-Russell (H-R) diagram, which correlates stellar luminosity with spectral class or temperature

1917
The 100-inch (2.5-m) Hooker Telescope is completed at Mount Wilson Observatory

1918
Harlow Shapley makes the first accurate estimate of the size of the Milky Way Galaxy

1919
Edwin Hubble begins observing with the Hooker telescope at Mount Wilson Observatory

1920
Vesto Slipher announces that most galaxies have red shifts, indicating that they are all rushing away from us

1923
Hubble discovers a Cepheid in the Great Nebula in Andromeda and establishes that it is a separate star system—an external galaxy

1929
Hubble connects the red shifts of galaxies to their speed of recession and suggests that the universe is expanding

1930
Clyde Tombaugh discovers a ninth planet— the diminutive Pluto

1931
Karl Jansky detects radio waves coming from the heavens, which leads to the foundation of radio astronomy

1948
The 200-inch (5-m) Hale Telescope is completed at Mount Palomar Observatory, near Los Angeles

Hermann Bondi and Thomas Gold put forward the steady-state theory for the origin and evolution of the universe, later developed by Fred Hoyle. The theory states that the universe has always existed in the state it is now and will exist for ever in that steady state

1955
The first giant, steerable radio telescope (250 feet [82 m] in diameter) is completed at Jodrell Bank Observatory, near Manchester, England

1957
Russia's satellite Sputnik 1 is launched on October 4, pioneering the Space Age

1958
The United States launches its first satellite, Explorer 1, on January 31; the satellite discovers the van Allen belts of radiation circling the Earth

1959
Russia's Luna 2 probe crashes on the Moon; Luna 3 returns the first pictures of the lunar far side, always hidden from the Earth

1961
Russian cosmonaut Yuri Gagarin becomes the first human to orbit the Earth on April 12, which he does once, in a Vostok capsule

1962
John Glenn becomes the first US astronaut in orbit, circling Earth three times on February 21 in the Mercury capsule "Friendship 7"

The United States launches the first successful deep space probe, Mariner 2, which makes a distant but effective flyby of Venus

1963
Maarten Schmidt first identifies a quasar, while studying radio source 3C-273; he identifies it with a starlike object that has an astonishingly large red shift, proving that it is very much father away than a star can be

1965
Mariner 4 probe sends back the first close-up photographs of a planet—Mars

1967
Cambridge radio astronomer Jocelyn Bell-Burnell discovers the first pulsar

1969 Apollo 11 astronauts Neil Armstrong and Edwin ("Buzz") Aldrin become the first humans to set foot on the Moon

1973
Launch of the US space station Skylab; three crews there carry out an in-depth study of the Sun

1974 Pioneer 10 probe observes Jupiter from close quarters

1976
Two Viking probes reach Mars; orbiters survey the planet from orbit; landers touch down on the surface, take photographs, report on Martian weather and test soil (in vain) for signs of life

1977
Voyagers 1 and 2 are launched to explore the outer planets

1979
Pioneer-Saturn flies by Saturn, taking the first close-up pictures of the ringed planet

1981
The US space shuttle makes its debut on April 12, when orbiter Columbia thunders into space

1986
Voyager 2 reports from Uranus; Giotto (from Europe), Vega 1 and 2 (from Russia) and Sakigake and Suisei (from Japan) target Halley's Comet

1989
Voyager 2 makes its final planetary encounter, with Neptune

1990
Orbiter Discovery launches the HST on April 24, as part of shuttle mission STS-31

1993
Astronauts on shuttle mission STS-61 in December repair the HST in orbit, correcting its defective vision with COSTAR and fitting a new wide field/planetary camera and solar panels

1997
The second HST servicing mission (STS-82) takes place in February, involving the installation of a new spectrograph (STIS) and a near-infrared camera and spectrometer (NICMOS)

1999
Astronauts replace the HST's attitude-maintaining gyroscopes on the HST servicing mission STS-103 in December

2002
Astronauts install an advanced digital camera (ACS) to improve the quality of the HST's images on servicing mission STS-109 in March

2003
The HST spies a giant radio jet coming from a spiral galaxy (0313-192); it is the first spiral galaxy known to emit jets

Shuttle orbiter Columbia breaks up just before landing on February 1, with the loss of the crew of seven. The three remaining shuttle orbiters are grounded and future missions put on hold

2010
The HST is scheduled to be at the end of its life; the next generation space telescope, named James Webb Space Telescope (JWST), should be ready for launch to replace it

Index

Acknowledgements

The author would like to extend his grateful thanks to the many individuals and organizations who have provided such invaluable assistance in the preparation of this book, particularly in the provision of images. Special mention must be made of Cheryl Gundy and the Public Outreach team at the Space Telescope Science Institute in Baltimore; Gwen Pitman and the photo research department at NASA headquarters in Washington D.C.; Mike Gentry and Gloria Sanchez of the Media Resource Center at the Johnson Space Center in Houston; Stephane Corvaja and Nadia Imbert-Vier at the European Space Agency in Paris; and Hans-Hermann Heyer of the European Southern Observatory's public relations department in Munich. Thank you all!

Picture Credits

(t=top, b=bottom, l=left, r=right, c=center)

All Hubble Space Telescope images are credited to NASA and the following additional teams and institutions.

ACS is an abbreviation for the Advanced Camera for Surveys team
AURA, the Associated Universities for Research in Astronomy
ESO, the European Southern Observatory
ESA, the European Space Agency
GFSC, the Goddard Spaceflight Center
HH, the Hubble Heritage team (STScI/AURA)
JPL, the Jet Propulsion Laboratory
JSC, the Johnson Space Center
NASA, the National Aeronautics and Space Administration
NRAO, the National Radio Astronomy Observatory
SP, Spacecharts Photolibrary
STScI, the Space Telescope Science Institute
U., University

Page 2 STScI, La Plata Observatory; 4 ESA, HH; 6 HH; 12 HH; 14 HH; 16 ACS, ESA; 19 Arizona State U., ESA; 20 ACS, ESA; 21 ESA, HH; 22 HH; 24b Harvard-Smithsonian Center, Steward Observatory, Rice U.; 25 Dr Robert O'Dell; 26t HH; 26b HH, U. of Hawaii; 27r HH; 29t STScI; 29b HH; 29c U. of Colorado; 29b Caltech, JPL; 30, 31 Arizona State U.; 32 ESA, Observatoire de Paris; 33 HH; 34t U. of Montreal, U. of Strasburg; 34b John Hopkins U.; 35 STScI; 36t Georgia Southern U., Cornell U.; 36b STScI; 37 HH; 38 HH; 39tb STScI; 40 HH; 43 U. of Illinois, U. of Washington; 44 U. of Arizona, NICMOS team; 45b Rice U.; 46 HH; 47 HH; 48 U. of Maryland; 49l STScI; 49r Instituto de Astrofisica de Canarias, Osservatorio Astronomico di Padova-Asiago, Nordic Optical Telescope; 50 HH; 52 JPL; 55 HH; 57 STScI; 58 U. of Colorado; 60b STScI; 61 HH; 62 ESA, STScI; 63r Arizona State U.; 64 HH; 69r STScI; 70 HH; 71l Carnegie Observatories; 71r STScI;

72 U. of Maryland, Caltech, Sabaru Telescope, U. of Leicester; 73t HH; 73b Tel-Aviv U., Columbia U.; 74 HH; 77 STScI; 78t NRAO; 78b STScI; 79l Princeton U., U. of Wales, STScI; 79r STScI; 80t U. of Alabama, Gemini Observatory, NRAO; 80b Yale U.; 81 HH; 82b STScI; 84 ACS, ESA; 85b STScI; 86 ACS, ESA; 89 HH; 90 STScI; 92 K. Lanzetta; 93 STScI; 95 STScI; 97 STScI; 98 STScI; 99 STScI; 101 U. of Illinois; 102t GFSC, STScI; 102b Carnegie Observatories, Columbia U.; 104 HH; 105l STScI; 105r U. of Manitoba, Pennsylvania State U., U. of Arizona, STScI; 106t STScI; 106b ACS, Hebrew U., John Hopkins U., Lick Observatory; 107 U. of Hawaii, STScI; 108t Caltech, Observatoire Midi-Pyrenees, ESA; 109 HH; 110 ESA, Observatoire de Paris; 112 GSFC, STScI; 114 STScI; 115t Extrasolar Research Corp.; 117t Caltech, JPL; 117b New Nexico State U., STScI; 122t Applied Research Corp., STScI; 123tr STScI; 124, 125 HST Comet Science Team; 126 HH; 127l JPL; 127cb STScI; 130 Cornell U., U. of Arizona; 136t U. of Colorado, Cornell U., STScI; 136b U. of Colorado, U. of Toledo, STScI; 137r HH; 138 Cornell U., HH; 139t U. of Michigan; 139bl STScI; 140 HH; 141c STScI; 142t U. of Arizona; 142b Cornell U., U. of Arizona; 143t U. of Arizona; 143b JPL; 144t U. of Arizona; 145b U. of Arizona; 146r U. of Wisconsin-Madison; 147t Southwest Research Institute, Lowell Observatory, ESA; 147c ESA, ESO; 168b STScI; 169 HH; 170t HH; 170b U. of Colorado, U. of Toledo; 171t ACS; 171b STScI; 172br STScI; 176t STScI; 177 UCLA; 179l ACS, Hebrew U., John Hopkins U., Lick Observatory; 180 HH

Additional images are credited to the following sources:

9 JSC; 18 SP; 24t SP; 27l ESO; 31t SP; 41 SP; 45t SP; 53 ESO; 54 SP; 56 ESO; 59 SP; 60t SP; 63l SP; 66 SP; 67 ESO; 68l NRAO; 69l ESO; 76 SP; 82t NRAO; 82r ESO; 88bl SP; 88br NRAO; 94 SP; 96b SP; 100 SP; 103 SP; 108b SP; 113 ESO; 116 SP; 118l ESA; 118r, 119 SP; 120, 121 SP; 122b Martin Moberley; 123tl Robin Kerrod; 123b SP; 127tr SP; 128, 129 SP; 132, 133 SP; 134, 135 SP; 137l SP; 139br SP; 141tb SP; 144b SP; 143tc SP; 146l SP; 147b SP; 148 ESO; 150 Robin Kerrod; 151 ESO; 152 Robin Kerrod; 152t NRAO; 153bl NRAO; 153br SP; 154, 155 SP; 156 SP; 157t Robin Kerrod; 157b SP; 158, 159 SP; 160, 161 SP; 163 JSC; 164 SP; 165 JSC; 167 SP; 168t SP; 172 JSC; 172bl JSC; 173 JSC; 174, 175 JSC; 176b JSC; 178 JSC; 179r JSC;